SUPERCAPACITOR DESIGN AND APPLICATIONS

Edited by **Zoran Stevic**

Supercapacitor Design and Applications

http://dx.doi.org/10.5772/62613
Edited by Zoran Stevic

Contributors

Jian-Gan Wang, Dewei Chu, Fadhel El Kamel, Kang Sun, Yue Zhou, Masatoshi Uno, Azrul Azlan Hamzah, Hafzaliza Erny Zainal Abidin, Jumril Yunas, Burhanuddin Yeop Majlis, Mohd Ambri Mohamed, Hengzhao Yang, Ying Zhang, Can Liu

CBS, Edition **2017**

Published by InTech

Janeza Trdine 9, 51000 Rijeka, Croatia

Publishing Process Manager Maja Bozicevic

Technical Editor SPi Global

Cover Designer InTech Design Team

Additional hard copies can be obtained from orders@intechopen.com

Supercapacitor Design and Applications, Edited by Zoran Stevic
p. cm.
Print ISBN 978-953-51-2748-2
Online ISBN 978-953-51-2749-9

Contents

Preface

Standard systems for electrical energy storage are capacitors and rechargeable batteries. Recently, more attention was paid to supercapacitors as a qualitatively new type of capacitors and fuel cells as sources of electricity. With the development of new materials and technologies, a very large surface area and very small interelectrode distances have been reached. This equates to an extremely high capacitance (several orders of magnitude greater than conventional capacitors), so such systems are called supercapacitors.

A lot of teams and laboratories around the world are working on the development of supercapacitors, while their better and better performance enables wider usage. The goal of this book is to bring closer to the readers new supercapacitor technologies that are changing the present and the future of electricity storage.

The book is divided into two sections. The first section deals with the current status and trends of development of supercapacitors, while the second section deals with applications of supercapacitors.

Prof. Dr. Zoran Stevic
University of Belgrade
Technical Faculty Bor
Belgrade, Serbia

Section 1

Supercapacitor Design

Chapter 1

High Volumetric Performance Supercapacitors with Controlled Nanomorphology

Yue Zhou and Qiming Zhang

Additional information is available at the end of the chapter

http://dx.doi.org/10.5772/65186

Abstract

Supercapacitor is one of the promising energy storage devices due to its relatively higher energy density compared with dielectric capacitor and higher power density and longer cycle life time (>millions) than conventional battery. In order to satisfy various requirements for energy technologies, supercapacitors with higher energy and power densities are required. In this chapter, we improved the electrochemical performance largely compared with commercial product through controlling the nanomorphology of cells. Meanwhile, although many past research programs have focused mainly on gravimetric energy densities, here we have also devoted efforts to study and develop nanomorphologic structures to realize high volumetric energy and power densities, since device volume is another critical and key performance parameter. Moreover, fundamental studies have been carried out on the mobile ion transport and storage in the nanostructures developed in this chapter.

Keywords: supercapacitor, carbon nanotubes, graphene, volumetric performance, nanomorphology control

1. Introduction

In order to meet different application requirements and also for fundamental studies of ion transport and storage in nanoporous media, we selected carbon-based electrodes with unique and controlled nanomorphologies: highly aligned carbon nanotube (A-CNT) forests. As synthesized, A-CNTs have low volume fraction of CNT (~1%). Traditional method to achieve high volumetric performance of A-CNTs was to employ liquid surface tension densification, which did not have control on the nanomorphology of A-CNTs after densification. In this work, making use of and improving upon the mechanical densification method, we achieved 40% volumetric density of A-CNTs. A series of supercapacitor electrodes have been developed and characterized. As can be seen in the section, the high volume fraction of A-CNTs and highly aligned ion channels in the nanoporous electrodes lead to the superior performance of the supercapacitors compared with any CNT-based supercapacitor electrodes studied earlier. The supercapacitors exhibit a volumetric power density, 25 kW/L (and gravimetric power density 50 kW/kg) for the capacitor cell with 0.8-mm thick A-CNTs, compared with the similar capacitors using A-CNTs densified by the liquid collapsing method, 13.4 kW/L (24 kW/kg) for cells with 0.5-mm thick A-CNTs. The study also shows the importance of the ionic conductivity of electrolytes in controlling the power and energy densities of the supercapacitors.

Since the energy and power densities of supercapacitors are directly proportional to the square of cell operation voltage V (~ V^2), raising the cell operation voltage will have great potential to enhance the energy and power densities. Asymmetric supercapacitors, which allow for optimization of both cathode and anode simultaneously, provide an attractive approach to raise the cell operation voltage, besides other properties. In this chapter, we have investigated asymmetric supercapacitor configurations for carbon-based electrodes for one electrode and conducting polymer (CP)-coated A-CNTs for the other one, based on their electrochemical windows. Here, we investigate the asymmetric supercapacitors where both electrodes are tailored, respectively, to improve the device electrochemical performances such as specific capacitance and the electrochemical window. Hence, operating voltage is increased. The conformal vapor is utilized to deposit CP on the A-CNTs, enhancing the charge storage capability of the electrode, while the aligned nanowire morphology of the composite electrode exhibits straight fast ion transport pathways to enhance power. The a-graphene electrode, which is fabricated through a self-assembly process, shows the high active material density. Combining with a high specific surface area of 3000 m^2/g, the electrode yields very high specific volumetric capacitance, energy, and power densities. As a result, the asymmetric supercapacitors show an energy density 113 Wh/L (176 Wh/kg), which is the highest among all carbon-based supercapacitors, and a power density 149 kW/L (233 kW/kg).

2. Symmetric supercapacitors with controlled unique nanomorphology

Presently, most supercapacitors are fabricated from activated carbon (AC), which possesses a very large specific surface area (1000–2000 m^2/g). Recent advances have demonstrated many attractive features of utilizing A-CNTs for supercapacitors with nanoporous electrodes, especially the parallel ion channels formed by the A-CNTs that improve the ion transport, as schematically illustrated in **Figure 1(a)**, compared with randomly arranged nanoporous electrodes from AC, forming tortuous ion transport pathways [1, 2]. Consequently, supercapacitor cells with A-CNTs exhibit higher power and energy density than that from AC. Since as grown A-CNT forests have CNT volumetric density <5 vol%, A-CNTs should be densified to reach higher A-CNT volumetric density for practical supercapacitor applications. In the past decade, many works have been conducted to produce aligned A-CNTs with high CNT density to achieve high volumetric capacitance, energy density, and power density, which are critical for modern electric and electronic systems to realize compact device size and increased functions within given device volumes. For example, Futaba et al. [3] employed a liquid collapsing method to densify A-CNTs with a high density (~50 vol%). Here, the mechanical densification method has several advantages compared with the liquid collapsing method. This method provides a precise control on the density of the final A-CNTs, ranging from the original 1 vol% A-CNTs to >50 vol%. Besides, A-CNT samples with different sizes can be densified with precisely controlled nanomorphology (alignment), and hence this method provides a realistic pathway for scaling up high-density A-CNTs for large-scale manufacturing of supercapacitors from A-CNTs of ultrahigh volume density [4]. In addition, the availability of A-CNTs with different density and hence different ion channel sizes also creates a great opportunity to study how the ion channel size formed by the A-CNTs affect ion transport and storage, which is of great importance in developing supercapacitors with high energy, power density, and tailored performance. We have also studied the influence of the ionic conductivity of electrolyte on the electrochemical performance of the supercapacitors. An imidazolium-based ionic liquid (1-ethyl-3-methylimidazolium-tetrafluoroborate, EMI-BF_4) was chosen for the study. Imidazolium ionic liquids (ILs) due to their high ionic conductivity and wide elec-

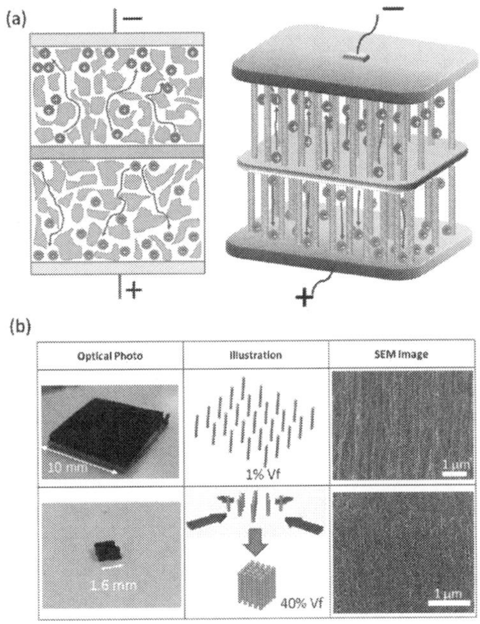

Figure 1. (a) The illustration of the tortuous ion transport paths in the activated carbon electrodes as well as parallel ion pathways in the A-CNTs. (b) Optic images, mechanical densification process, and SEM images of 1% Vf and 40% Vf A-CNTs.

trochemical window have been investigated very extensively as the electrolytes for supercapacitors. A mixture of an IL such as EMI-BF_4 and molecular liquid such as propylene carbonate (PC) can lead to marked enhancement (more than two times higher) in the ionic conductivity compared with the pure EMI-BF_4. The experimental results show that increased ionic conductivity of the electrolyte can lead to a large increase in the power density (more than double the power density using EMI-BF_4/PC compared with pure EMI-BF_4) of the supercapacitors.

The A-CNTs in this work were fabricated through a modified chemical vapor deposition (CVD) method on silicon wafers and iron (Fe) on alumina was used as the catalyst. The as-grown carbon nanotube forests have a 1% volume fraction (Vf) with density of 10^9–10^{10} CNTs cm^{-2}. The average diameter of nanotubes is 8 nm with 3–5 multiwalls. The spacing between nanotubes is approximately 80 nm in the forest. For the high Vf A-CNT synthesis, the freestanding CNT forests were released from the silicon wafer using a razor blade. And then as shown in **Figure 1(b)**, the forests were subjected to a mechanical biaxial densification process in two orthogonal directions. In this method, the A-CNT forest was densified along one direction firstly to a fixed distance by utilizing a mechanical bar, and then another mechanical bar in the orthogonal direction was employed to compress the A-CNT forest to the final volume fraction. By varying the inter-CNT distance in densification process, A-CNT forests with different Vfs can be obtained.

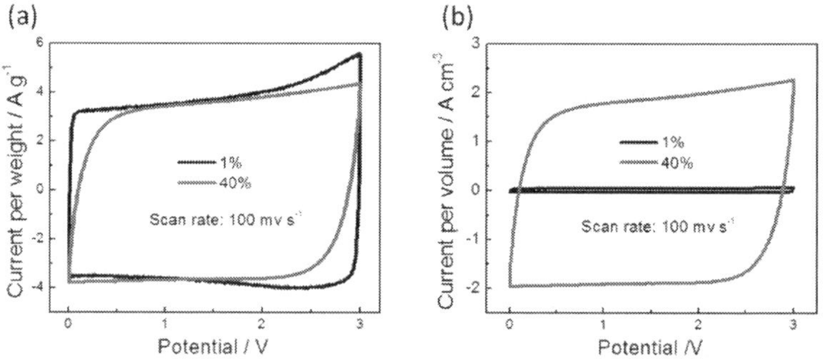

Figure 2. Electrochemical performance of supercapacitors with A-CNT electrodes and EMI-BF4/PC electrolyte at 4 V: (a) gravimetric cyclic voltammograms and (b) volumetric cyclic voltammograms of A-CNT electrodes with 1% and 40% Vf at 100 mV s^{-1}.

The A-CNT forests were used as the electrodes of the supercapacitors. 3 M EMI-BF_4 (1-ethyl-3-methylimidazolium tetrafluoroborate) in propylene carbonate (PC) was used as the electrolyte to improve ionic conductivity. Compared with pure ionic liquid, the ionic liquids/molecular liquids (IL/ML) mixture will have higher conductivity. Polypropylene porous membrane (Celgard 3501, Celgrad LLC) with 25 μm thickness was used as the separator. The sandwich architecture (electrode/separator/electrode) was placed between two pieces of Au-coated steel plates, which served as the current collectors. Then, the whole cell was housed in a Teflon holder. As a comparison, supercapacitor electrodes, which are made from the activated carbon powders mixing with 10 wt% PTFE and 10 wt% carbon black, were also fabricated using the standard method.

Figure 2(a) presents the cyclic voltammetry (CV) curves at 100 mV s^{-1} scan rate of supercapacitors with 1% Vf and 40% Vf of A-CNTs as electrodes in 3 M EMI-BF_4/PC electrolyte. As shown in the figure, there is very little difference in the gravimetric capacitance drop as the A-CNT Vf increases from 1% to 40%, showing that the densification process can still maintains the aligned nanomorphology with parallel ion pathways. On the other hand, increasing the A-CNT Vf from 1% to 40% leads to a large increase considering the volumetric capacitance performance, as shown in **Figure 2(b)**.

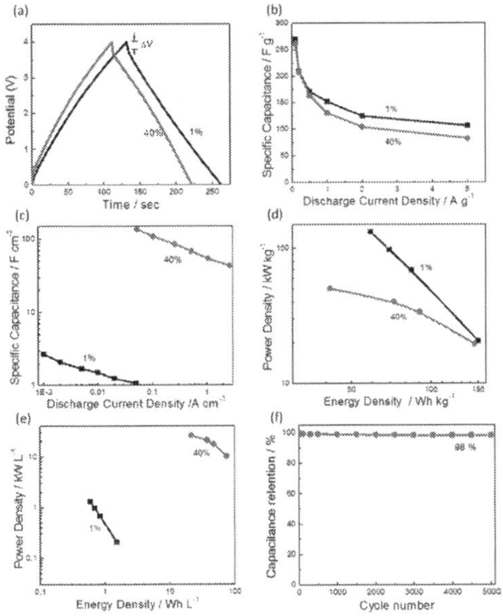

Figure 3. (a) Galvanostatic charge and discharge curves at 1 A g^{-1} for supercapacitors with electrodes of 1% and 40% Vf forests and EMI-BF4/PC as electrolyte. (b) Gravimetric and (c) volumetric specific capacitances for the supercapacitors with electrodes of 1% and 40% Vf forests with different discharge rates. (d) Gravimetric and (e) volumetric Ragone plots for the supercapacitors with electrodes of 1% and 40% Vf A-CNTs. (f) Cycle retention performance of 40% Vf A-CNT supercapacitor with a voltage of 4 V under the charge and discharge current density of 5 A g^{-1}.

The galvanostatic cycles between 0 and 4 V for the supercapacitors with 1% Vf and 40% VfA-CNT electrodes under 1 A g^{-1} current density are presented in **Figure 3(a)**. The capacitance of the cell can be determined:

$$C = I / \left(dV / dt \right) \quad (1)$$

where I is the constant current density, V is potential, and t is discharge time. **Figure 3(b)** and **3(c)** shows the specific gravimetric and volumetric capacitances with different discharge currents, respectively. Although the cell with electrodes of 1% Vf show a little higher specific gravimetric capacitance which is larger than 270 F g^{-1}, their volumetric capacitance is very low (around 3 F cm^{-3}).

It should be mentioned that the 1% Vf A-CNT electrode has a very low active material density with 0.013 g/cm^3. For this kind of electrode, the majority of the electrode volume is filled with electrolytes, whose mass is not included when evaluating the gravimetric electrochemical performance. Instead, the volumetric values should be utilized to investigate when comparing electrodes with large difference about the active material density. It could be found that the specific volumetric capacitance of 40% Vf A-CNT electrodes is about 40 times higher than that of 1% Vf A-CNT electrodes, exhibiting that the nanomorphology of the A-CNTs is preserved by utilizing the mechanical densification method developed here.

The maximum power density and energy density of the supercapacitor cells can be calculated based on the equivalent series resistance (ESR) and the specific capacitance. The gravimetric and volumetric Ragone plots for the supercapacitor cells are shown in **Figure 3(d)** and **3(e)**, respectively. The energy of the supercapacitor cell under each discharge current can be calculated by integrating the discharge curves with time.

$$E = \int IV(t)dt \tag{2}$$

The ESR can be calculated based on the equation below:

$$ESR = \Delta V / \Delta I \tag{3}$$

where ΔV is the voltage drop as the current is switched from a positive value to a negative value, such as from 1 Ag^{-1} to −1 Ag^{-1} ($\Delta I = 2$ Ag^{-1}). The maximum power density, hence, can be deduced from:

$$P = V^2 / (4 \times ESR) \tag{4}$$

The active material density of the electrodes in many recently developed nanoporous electrodes, such as the A-CNT electrodes described here, can vary over a broad range. The traditional method to evaluate the performance of supercapacitor cell such as the gravimetric energy and power densities will not accurately reflect the device performance because the only mass of the conductive electrode material is included and the active material density is usually very low. For example, due to a very large "empty space" (80 nm between nanotubes) in the ion pathway, a very high power density of 100 kW kg^{-1} was obtained in the supercapacitor cell with 1% Vf A-CNT forests as the cell electrodes. As a comparison, 50 kW kg^{-1} was obtained in the supercapacitor cell based on 40% Vf A-CNT forests as electrodes due to the smaller pore size. However, the supercapacitor cell with 1% Vf A-CNT forests shows a very low gravimetric maximum power and energy density compared with that of the supercapacitor cell based on 40% Vf A-CNT forests if the total electrode mass, including all the elements such as the active materials and the electrolytes, is considered.

For the supercapacitor cells with 1% Vf of A-CNT forests as electrodes, the specific gravimetric capacitance will decrease to 4.3 Fg^{-1} when the total electrode mass, including both the 1% Vf active material (A-CNTs) and 99% Vf electrolyte, is used for the calculation. This is much smaller than 270 F g^{-1} calculated when only the mass of the A-CNTs is included. In contrast, for the 40% Vf A-CNTs, a gravimetric capacitance of 270 F g^{-1} for the active material alone is equivalent to a 139.8 F g^{-1} when all the electrode mass is included.

The supercapacitors based on 40% Vf A-CNTs also exhibit an excellent cycling life as shown in **Figure 3(f)**. The data were acquired over 5000 cycles by repeating the galvanostatic charge and discharge process between 0 and 4 V under an alternate current densities of 5 and −5 A g^{-1}, which show an excellent electrochemical stability. Capacitance retention of 98% after 5000 cycles was obtained based on the supercapacitor cell with the ultra-high density A-CNTs as 40% Vf.

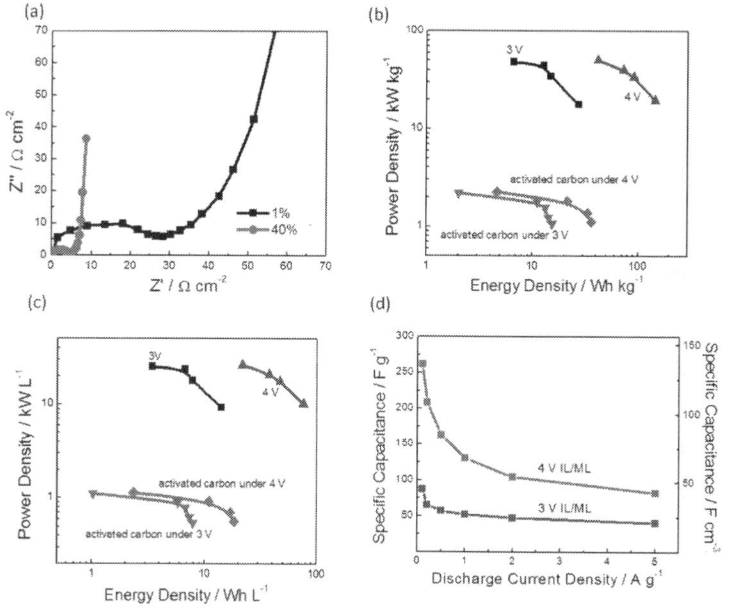

Figure 4. (a) Nyquist plots of supercapacitors based on A-CNTs with the two volume fractions in the range of 100 kHz to 10 mHz. (b) Gravimetric and (c) volumetric Ragone plots of supercapacitor cells with electrodes based on 40% Vf A-CNT forests and with electrodes using activated carbon. (d) Under several discharge current densities, the relationship of specific capacitance of cells based on 40%Vf A-CNT forests.

The electrochemical impedance spectroscopy (EIS) analysis was carried out to investigate the possible influence of the ultra-high density A-CNT forests on the electrochemical performance of supercapacitor electrodes. **Figure 4(a)** shows the EIS figure in the frequency range of 100 kHz–10 mHz. The Nyquist plots of supercapacitor cells with electrodes based on 1% and 40% Vf A-CNT forests show both semicircles in the high/middle frequency and sharp rises at low frequency range. The semicircle behavior is due to the charge transfer resistance of the electrodes, and the sharp increase at low frequency range is resulted from the ideal capacitive performance of the electrode. It could be found that the cell with electrodes based on 40% Vf A-CNT forests also show much smaller resistance (Z′) when normalized with the area of the capacitors (Ω cm^{-2}) compared with that of electrodes based on 1% Vf A-CNT forests.

As a comparison, activated carbon with the thickness of 800 μm was fabricated. The maximum power density, energy density, and electrochemical performances of cells based on activated carbon are shown in **Figure 4(b)** and **4(c)**. A much lower volumetric energy density and power density (20 and 1.1 kWL^{-1}, under 4 V) were obtained based on the activated carbon electrodes, compared with the performance with 40% Vf of A-CNT forests (75 and >25 kW L^{-1} under 4 V). These results indicate the superior electrochemical performance of the ultra-high-density A-CNT electrodes fabricated from the mechanical densification method developed here. The nanomorphology of the aligned ion pathways leads to the fast charge/discharge rate and high power/energy densities.

Figure 4(b) and **4(c)** presents that there is very little increase in the power density while increasing the voltage from 3 to 4 V leads to a large increase in the energy density, from 15 to 75 Wh L^{-1} (gravimetric 30–150 Wh kg^{-1}). The large increase in the energy density results from the increase in the specific capacitance of the electrodes with the increasing of voltage, as shown in **Figure 4(d)**. The specific capacitance is obtained as 260 F g^{-1} (135 F cm^{-3}) at 4 V. These values are much higher than those reported earlier for the supercapacitors utilizing densified A-CNTs through the liquid collapsing method. On the other hand, it is noted that the maximum power density depends on the ESR (see Equation (4)), as well as the applied voltage. The results exhibit that there is a large increase in the ESR as the operation voltage has increased from 3 to 4 V, which is consistent with the results of an earlier study in our group [5]. Diffusion process and drifting process have dominated the transport of mobile ion in ionic devices such as supercapacitors. Diffusion is relatively slower and hence represents much higher ESR compared with drifting process.

3. Asymmetric supercapacitor with high electrochemical performance

In this work, an asymmetric supercapacitor, exploiting nm-scale conformal coating of a conducting polymer, poly(ethylenedioxythiophene) (PEDOT) on aligned carbon nanotubes (A-CNTs) as one electrode and an ultra-high density aligned graphene sheets (a-graphene) as the other electrode, has been developed. The asymmetric configuration of the supercapacitor allows both electrodes to be separately tailored, increasing device capacitance and the electrochemical window, and thereby operating voltage. As a result of complementary three-dimensional nanotailoring of the asymmetric electrodes, the device exhibits a wide 4V electrochemical window and high electrochemical performance [6].

For supercapacitors, it is well known that the energy density (*E*) is related to the gravimetric or volumetric cell capacitance (*C*) and operation voltage (*V*), i.e.

$$E = \frac{1}{2}CV^2 \tag{5}$$

And the maximum power density P is determined by Equation (4). Equations (4) and (5) have shown that one of the most effective ways to increase both the power and energy densities is to raise the cell operation voltage. In general, the operation voltage of supercapacitor cell has the relationship with the electrochemical window, which is determined by the interface between the electrode and electrolyte. As a promising way, asymmetric supercapacitor can be assembled to make full use of the electrochemical windows of both electrodes to increase the maximum cell operation voltage in the supercapacitor cell. Morphology control of the electrodes via nanoscale tailoring is shown to be an effective way to increase supercapacitor performance (gravimetric and volumetric power and energy) via increasing ECW and capacitance and reducing ESR.

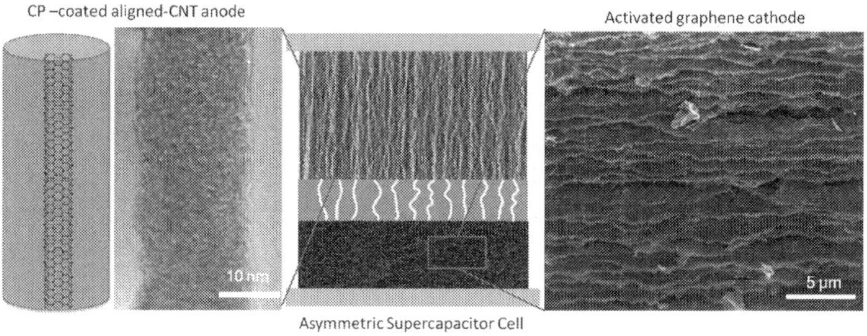

Figure 5. Nanostructured electrodes in asymmetric supercapacitors. Left, Low and high magnification TEM micrographs of the electrode, composed of conformal oCVD PEDOT on A-CNTs, and right, SEM images of activated graphene electrode.

Recent advances in the conformal coating of conducting polymer PEDOT by oxidative chemical vapor deposition (oCVD) onto nanowire arrays and development of graphene with relatively high gravimetric surface area create unique opportunities for developing high performance asymmetric supercapacitors. As schematically illustrated in **Figure 5**, the combination of the aligned ion transport pathways formed by the aligned nanowire arrays that provide fast ion transport in the electrode that reduces ESR of the electrode and the conformal coating of conducting polymer PEDOT on the A-CNTs that enhances the charge storage capability (a large C) contributes to both high energy and power densities of the cells. PEDOT was selected as the conducting polymer because of its environmental stability, high electrical conductivity, and a wide ECW. As shown in **Figure 6(a)**, conformally coated oCVD PEDOT on the A-CNTs yields a stable ECW from −1.0 to 1.8 V, when using an ionic liquid/molecular liquid mixture, 2 M 1-butyl-3-methylimidazolium tetrafluoroborate (BMI BF_4)/propylene carbonate (PC), as the electrolyte. The high ECW of 1.8 V makes it as an excellent positive electrode material in the asymmetric supercapacitors.

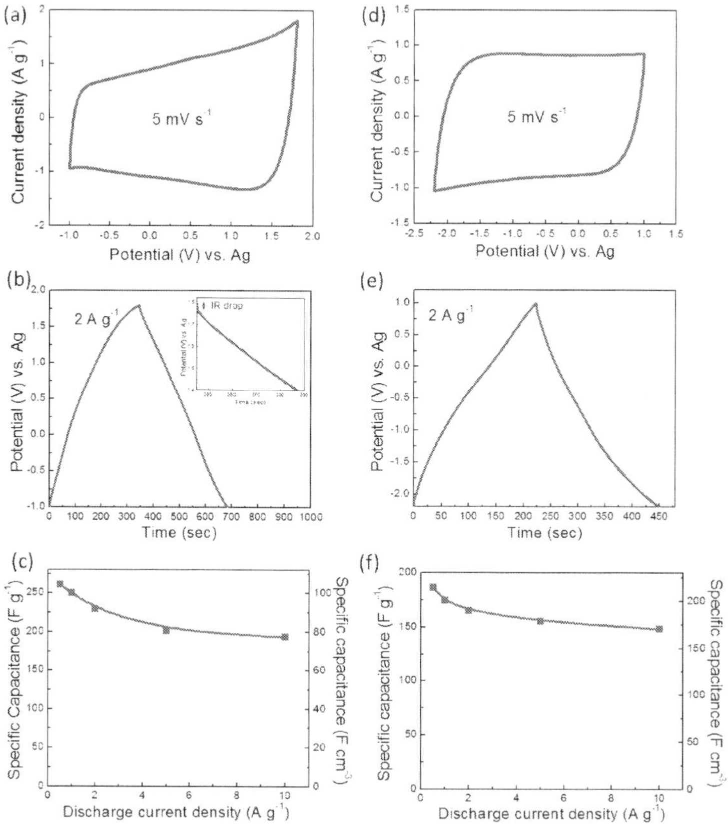

Figure 6. Performance of two electrodes: (a) CV curves of PEDOT/A-CNTs composite at 5 mV s^{-1} in 2 M BMIBF4/PC. (b) Galvanostatic charge/discharge curves of PEDOT/A-CNT composite at current densities of 2 A/g. (c) Specific capacitance at different discharge densities of PEDOT/A-CNT electrode. (d) CV curves of a-graphene at 5 mV/s in 2 M BMIBF4/PC. (e) Galvanostatic charge/discharge curves of a-graphene at current densities of 2 A/g. (f) Specific capacitance at different discharge densities of a-graphene electrode.

Due to their favorable ECWs, carbon-based electrodes such as activated carbon have been used for the negative in the asymmetric supercapacitors. In this study, a new class of carbon material, the a-graphene, was selected due to its superior gravimetric surface area compared with activated carbon. The a-graphene, first reported by Zhu et al., presented a very large specific surface area (as large as 3100 m^2/g) with nanosized pores and demonstrated a gravimetric capacitance as high as 200 F/g when assembling supercapacitors [7]. However, the simple mechanical packing of the a-graphene flakes led to a relatively low density (~0.3 g/cm^3) compared with the graphite density of 2.2 g/cm^3. A low volumetric efficiency of supercapacitors was obtained by this configuration (the specific volumetric capacitance was 60 F/cm^3). When randomly packing these a-graphene flakes, which have lateral dimension of a few microns and a thickness of a few nanometers, it is inevitable to include micron-sized pores in the electrodes, reducing the density. Self-assembly processes are quite effective in increasing the density of graphene-based materials. Here, by employing a vacuum-assisted self-assembly process, which enabled a-graphene flakes aligned in parallel and stacked successively on top of each other, as shown by the SEM image of **Figure 5**, we fabricated the a-graphene electrodes with high density while preserving the nanoporous morphology of each a-graphene flake. The ECW of the a-graphene was also characterized, and as presented in **Figure 6(d)**, the a-graphene has a stable ECW from −2.2 V to 1 V with an electrolyte of 2M $BMIBF_4/PC$. The combination of high specific gravimetric surface area and high density of the a-graphenes as the negative electrode increases the ECW and results in high volumetric power and energy densities, besides the long cycle lifetime and high capacitive retention.

The high density (through packing) of aligned carbon nanotube (A-CNT) forests are distinctively advantageous as the conductive networks to support the CP coating layer in supercapacitors, when we compare with randomly packed morphologies. Besides the direct (and thereby fast) ion transport in aligned channel to reduce ESR illustrated in **Figure 5**, the PEDOT/A-CNT composite also provides better mechanical stability and hence higher retention property after many charge and discharge cycles, compared with the electrodes of the PEDOT/randomly packed CNT networks. In the extant literature, most electrodes of CP/CNTs were fabricated by electrochemical methods, which will lead to nonuniform CP layers coating on the CNTs. As shown in **Figure 6**, the thin (~5 to 10 nm) oxidative chemical vapor deposition (oCVD), PEDOT layers form a conformal coating on very high aspect ratio A-CNTs (0.2 mm long).

The electrochemical performance of PEDOT/A-CNT composite electrode were investigated by cyclic voltammetry (CV) and galvanostatic charge-discharge tests using a screen-printed electrode system (Dropsens) with the PEDOT/A-CNT composite as the working electrode, while Ag and Pt were employed as the reference and counter electrodes, respectively. **Figure 6(a)** shows a CV curve of the PEDOT/A-CNT forest composite electrode in 2 M $BMIBF_4/PC$ electrolyte under a scan rate of 5 mV s^{-1}, which shows an ECW from −1 V to +1.8 V. The galvanostatic cycles for the PEDOT/A-CNT electrode at the alternate currents of 2 and −2A g^{-1} are shown in **Figure 6(b)**. The symmetric and linear charge and discharge characteristics with time reveal a rapid I-V response and reversible electrochemical reaction, leading to an superior capacitive behavior. The specific capacitance of the electrode can be determined from Equation (1). A high specific gravimetric capacitance of 230 F/g was obtained at 2 A/g. **Figure 6(c)** presents the specific capacitance at different discharge current density, from 0.5 to 10 A/g. Capacitance retention of 74.2% was obtained at 10 A/g (from 260.8 F/g at 0.5 A/g to 193.5 F/g at 10 A/g), indicating that the PEDOT/A-CNT electrode provide reliable capacitive performance for high power applications. This relatively high retention mechanistically arises from the conformal coating of oCVD PEDOT on A-CNTs. The parallel ion transport pathways formed by the PEDOT/A-CNTs and the high electronic conductivity of the A-CNTs improve the ion transport and result in low ESR and therefore high power density. The cycling stability of the PEDOT/A-CNT electrodes was characterized and compared with that of the electrodes of PEDOT deposited on randomly packed CNT networks. Symmetric supercapacitors made of the PEDOT/A-CNTs had a retention of 89% after 1000 cycles of 2 V voltage cycle compared with a retention of 73% after 1000 cycles from PEDOT on random CNT morphologies. In the randomly packed CNT networks, there are CP layers in the gaps between nanotubes. The mechanical failure of CP layers in these gaps will cause disruption of the electric conduction paths between nanotubes and reduce the conductivity of CNT networks after long charge/discharge cycles. As a result, the capacitance and other electrochemical performances will be influenced. In contrast, the electric conduction path of the continuously aligned CNT forests would not be disrupted by the mechanical failure of the CP coating layers due to this nanomorphology. Hence, the A-PEDOT/A-CNT electrodes exhibit more robust mechanical stability and high retention of the capacitance, compared with the electrodes of the CP deposited on randomly packed CNT networks.

The electrochemical performance of the a-graphene electrode was characterized as above, including using 2 M $BMIBF_4$/PC as the electrolyte. **Figure 6(d)** presents a CV curve of the a-graphene electrode at a scan rate of 5 mV/s, showing an ECW of −2.2 V to +1 V. The slope of the discharge curve in **Figure 6(e)** yields a specific capacitance of 165 F/g at 2 A/g. The specific capacitances of the a-graphene electrode with different discharge currents are presented in **Figure 6(f)**. The a-graphene electrode exhibits high specific capacitance, ranging from 186.4 to 148.2 F/g as the discharge current increases from 0.5 A/g to 10 A/g. Moreover, the high density of the a-graphene electrodes results here in a specific volumetric capacitance with 175 F/cm^3 from the discharge curve of constant current of 1 A/g, which is the highest among all the carbon-based electrodes.

Both electrodes are independently tailored in asymmetric supercapacitors to operate under more optimal conditions. Here, the PEDOT/A-CNTs electrode and a-graphene electrode were assembled, separated by a porous paper of 40 μm thick. 2 M BMIBF4/PC was used as the electrolyte due to its high ionic conductivity. By properly tuning the mass ratio of the two electrodes, the asymmetric capacitor can be operated at the full 4 V cell operation voltage, reaching the maximum voltages from both electrodes (=1.8 V (A-CNT/PEDOT composite, V_+) + 2.2 V (a-graphene flakes, V_-)). Based on the consideration that charge stored at the two electrodes should be equal in magnitude with opposite sign ($|q_+| = |q_-|$), the mass ratio between the two electrodes can be determined as following equation if we consider the stored charge q at the electrode is $q = C\Delta Vm$, where C is the specific gravimetric capacitance, ΔV is the maximum potential range allowed by the ECW, and m is the mass of the electrode.

$$\frac{m_+}{m_-} = \frac{C_-\Delta V_-}{C_+\Delta V_+} \tag{6}$$

http://dx.doi.org/10.5772/65186

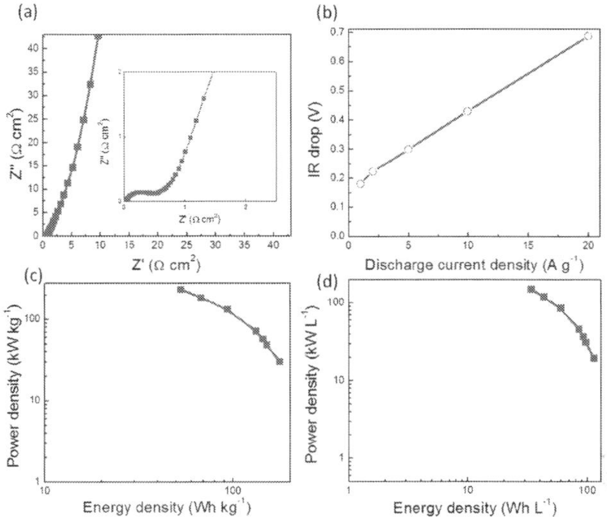

Figure 7. Cell performance: (a) CV curves of asymmetric cell under different scan rates from 5 to 100 mV/s between 0 and 4 V using 2 M BMIBF4/PC as the electrolyte. (b) Galvanostatic charge/discharge curves of asymmetric cell under a current density of 2 A/g. (c) Cell capacitances of asymmetric cell at different discharge current densities. (d) Cycle capacitance retention of asymmetric supercapacitor under a voltage of 4 V at a current density of 5 A/g in 2 M BMIBF4/PC electrolyte.

From the specific capacitances of the two electrodes, 230 and 165 F/g, respectively, under a constant discharge current of 2 A/g, and $\Delta V_+ = 1.8$ V and $\Delta V_- = -2.2$ V, Equation (6) can lead to the mass ratio (m_+/m_-) of 0.88, for a full 4 volts cell operation voltage, providing the design characteristics of the asymmetric supercapacitor assembled here.

Figure 7(a) presents the CV curves of the fabricated asymmetric supercapacitors at scan rates from 5 to 100 mV s^{-1} using the 2 M $BMIBF_4$/PC as electrolyte. The capacitors display near rectangular CV curves, especially for the lower scan rates. In order to evaluate the capacitive performance of the cell further, galvanostatic charge/discharge curves at different current densities were characterized. The galvanostatic cycles at alternate charge/discharge current densities of 2 and –2 A/g are presented in **Figure 7(b)**, from which the cell capacitance was determined (Equation (1)). **Figure 7(c)** presents the cell gravimetric and volumetric capacitances at different discharge currents. It should be noted that the calculated cell capacitance was based on the total mass of the active materials (both positive and negative electrodes) because it is not meaningful to deduce the specific capacitance of a single electrode for the asymmetric supercapacitor. The cell capacitance is 81.6 F/g at 0.2 A/g and becomes 55.4 F/g as the discharge current density increases to 10 A/g, indicating relatively good capacitance retention. The cell capacitance obtained here is higher than that of a-graphene-based symmetric supercapacitors and other conducting polymer-based asymmetric supercapacitors. Cycling retention performance of the asymmetric supercapacitors was investigated by continuously performing the galvanostatic charge/discharge process between 0 and 4 V at the alternative current densities of 5 and −5 A/g for more than 1000 cycles. The asymmetric supercapacitor maintains electrochemical retention of 94% after 1000 cycles as presented in **Figure 7(d)**. The small capacitance loss is likely attributable to capacitance decay of the PEDOT/A-CNT electrode.

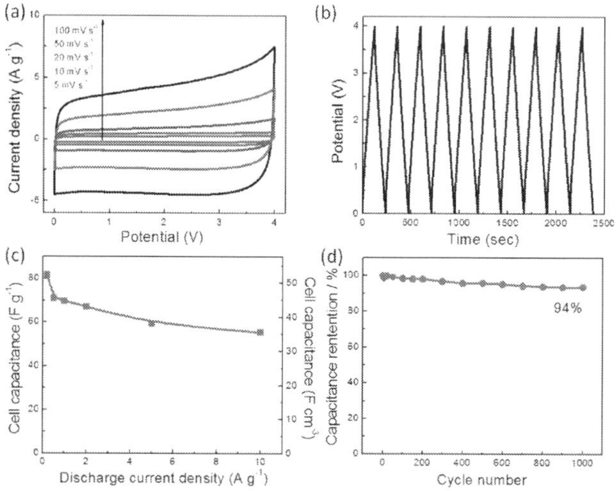

Figure 8. Asymmetric cell absolute and relative performance: (a) Nyquist plot of the asymmetric cell. (b) Internal resistance drop at different current densities. (c) Ragone plot of PEDOT/A-CNTs//a-graphene asymmetric supercapacitor in gravimetric unit. (d) Ragone plot of PEDOT/A-CNTs//a-graphene asymmetric supercapacitor in volumetric unit.

The electrochemical performance of the asymmetric supercapacitor cell was further characterized by electrochemical impedance spectroscopy (EIS). Nyquist plot, as shown in **Figure 8(a)**, is achieved in the frequency range of 100 kHz–10 mHz of 5 mV applied voltage. It could be found that a semicircle in the high-frequency region and a sharp rise of the imaginary part of the electric impedance shown in the figure reflecting the dominance of the cell capacitance in the-low frequency region. As shown in the Nyquist plot, the semicircle at the middle/high frequency is due to the charge transfer resistance in the porous electrodes. The high-frequency intersection on the real axis of the Nyquist plot shown in the figure represents the internal resistances. Internal resistance of 0.1 Ω cm^2 of the cell is obtained in the figure when normalized with the area of the current collector of the capacitors indicating a high electrical conductivity and low ESR of the cells.

The maximum power density of the asymmetric supercapacitor cell is determined from Equation (4). **Figure 8(c)** presents the Ragone plot (gravimetric power density versus energy density) of the asymmetric supercapacitor derived from the galvanostatic discharge curves measured at different charge/discharge current densities following standard practice. In addition to gravimetric performance, the maximum power density is derived from Equation (3), where V is the operation voltage, which is 4 V here.

The volumetric energy and power densities are more important in practical applications. The cells exhibit both high volumetric and gravimetric power and energy density at 149 kW/L (233 kW/kg) and 113.2 Wh/L (176.6 Wh/kg), respectively. These values are significantly higher than those of other reported carbon-based symmetric supercapacitors, conducting polymer-based supercapacitors and other devices reported previously [8–10]. The pseudocapacitor nature of PEDOT has lower charge/discharge speed compared with that of the pure EDLC supercapacitors and hence the power density is lower than that of the A-aMEGO supercapacitors, which have been presented in the proceeding section.

In summary, an asymmetric supercapacitor, employing the conformal coating of PEDOT/A-CNT composite as one electrode and high density a-graphene flakes as the other electrode, has been developed in this paper. PEDOT/A-CNT composite combines fast ion transport pathways, enhances charge storage capability, and reduces ESR while a-graphene electrode fabricated from a self-assembly process, which possesses exceedingly high specific gravimetric and volumetric capacitance. The two electrodes are individually tailored to control the nanomorphology and work synergistically together in the asymmetric cell configuration. The ECW has been expended up to 4 V. Tailoring of the two electrodes materials at a scale approaching that of the ions can allow asymmetric supercapacitor performance to be further expanded to meet the requirement of a broad range of energy storage applications.

Acknowledgements

This work was supported by AFOSR under Grant no. FA9550-11-1-0192. Reproduced from Ref. [4] with permission from ELSEVIER. Reproduced from Ref. [6] with permission from the Royal Society of Chemistry.

Author details

Yue Zhou* and Qiming Zhang

*Address all correspondence to: yuezhou26@gmail.com

Department of Electrical Engineering, Pennsylvania State University, University Park, Pennsylvania, USA

References

[1] W. Lu, L. Qu, K. Henry, L. Dai. High performance electrochemical capacitors from aligned carbon nanotube electrodes and ionic liquid electrolytes. Journal of Power Sources. 2009;189(2):1270. DOI: 10.1016/j.jpowsour.2009.01.009

[2] C.L. Pint, N.W. Nicholas, S. Xu, Z. Sun, J.M. Tour, H.K. Schmidt, R.G. Gordon, R.H. Hauge. Three dimensional solid-state supercapacitors from aligned single-walled carbon nanotube array templates. Carbon. 2011;49(14):4890. DOI: 10.1016/j.carbon.2011.07.011

[3] D.N. Futaba, K. Hata, T. Yamada, T. Hiraoka, Y. Hayamizu, Y. Kakudate, O. Tanaike, H. Hatori, M. Yumura, S. Iijima. Shape-engineerable and highly densely packed single-walled carbon nanotubes and their application as super-capacitor electrodes. Nature Materials. 2006;5:987. DOI: 10.1038/nmat1782

[4] Y. Zhou, M. Ghaffari, M. Lin, E.M. Parsons, Y. Liu, B.L. Wardle, Q.M. Zhang. High volumetric electrochemical performance of ultra-high density aligned carbon nanotube supercapacitors with controlled nanomorphology. Electrochimica Acta. 2013;111:608. DOI: 10.1016/j.electacta.2013.08.032

[5] J. Lin, Y. Liu, Q.M. Zhang. Charge dynamics and bending actuation in Aquivion membrane swelled with ionic liquids. Polymer. 2011;52(2):540. DOI: 10.1016/j.polymer.2010.11.030

[6] Y. Zhou, N. Lachman, M. Ghaffari, H. Xu, D. Bhattacharya, P. Fattahi, M.R. Abidian, S. Wu, K.K. Gleason, B.L. Wardle, Q.M. Zhang. A high performance hybrid asymmetric supercapacitor via nano-scale morphology control of graphene, conducting polymer, and carbon nanotube electrodes. Journal of Materials Chemistry A. 2014;2:9964. DOI: 10.1039/c4ta01785d

[7] Y. Zhu, S. Murali, M.D. Stoller, K.J. Ganesh, W. Cai, P.J. Ferreira, A. Pirkle, R.M. Wallace, K.A. Cychosz, M. Thommes, D. Su, E.A. Stach, R.S. Ruoff. Carbon-based supercapacitors produced by activation of graphene. Science. 2011;332(6037):1537. DOI: 10.1126/science.1200770

[8] B.G. Choi, M.H. Yang, W.H. Hong, J.W. Choi, and Y.S. Huh. 3D macroporous graphene frameworks for supercapacitors with high energy and power densities. ACS Nano. 2012;6:4020. DOI: 10.1021/nn3003345

[9] A. Izadi-Najafabadi, S. Yasuda, K. Kobashi, T. Yamada, D.N. Futaba, H. Hatori, M. Yumura, S. Iijima, K. Hata. Extracting the full potential of single-walled carbon nanotubes as durable supercapacitor electrodes operable at 4 V with high power and energy density. Advanced Energy Materials. 2010;22:E235. DOI: 10.1002/adma.200904349

[10] Z. Chen, J. Wen, C. Yan, L. Rice, H. Sohn, M. Shen, M. Cai, B. Dunn, Y. Lu. High-performance supercapacitors based on hierarchically porous graphite particles. Advanced Energy Materials. 2011;1:551. DOI: 10.1002/aenm.201100114

Chapter 2

Electrochemical Materials Design for Micro-Supercapacitors

Can Liu and Zhengjun Zhang

Additional information is available at the end of the chapter

http://dx.doi.org/10.5772/64986

Abstract

Micro–supercapacitors (m–SC) arise from the demand of developing micro–power system for MEMS devices, attracting much research interest in recent years. As m–SC has to achieve high areal energy and power densities, the volumetric capacitance and the rate capability of the electrode materials have become the most important concern. This review compares the intrinsic electrochemical properties of the state–of–art electrode materials for m–SC, reporting the recent advances in the three types of electrode materials. For carbon electrode materials, two developing trends are identified: one is to enhance volumetric capacitance through a proper film fabrication process, while the other one is to further promote its fast response rate by making open–structured devices. For pseudocapacitive oxides, in order to achieve better rate capability and cyclability, the relationship between the electrochemical property and the structure is worth further exploration. As an example, the composition, microstructure, and morphology of the molybdenum oxide film were optimized to realize superior electrochemical performance as an electrode material for m–SC. Architecture design is another important factor for m–SC. In–plane interdigital architectures have proven its success to fabricate fast response devices. Further study on the interplay effect between such architecture and pseudocapacitive materials is in need.

Keywords: micro-supercapacitors, electrode materials, EDLC, pseudocapacitance

1. Introduction

Since the microelectromechanical systems (MEMS) develop rapidly toward standalone microsensors, actuators, and various functional devices, the design of power supply has received more and more research interests [1]. The conventional bulky batteries severely limit the advantages of these smart systems, a micropower system, i.e., generating power directly from microstructures, thus has to be developed [2]. After the earliest explorations on the miro-internal-combustion engine which requires complicated micromachining processes and high manufacturing cost [2–4], researchers turn to microbattery system which is potentially of low cost and high capacity and more desirable for MEMS devices [1]. A complete micropower system should consist of energy conversion and storage units integrated on chip. The energy conversion devices include microscaled fuel cells and solar cells, while the energy storage devices mainly refer to rechargeable microbatteries, which have been remarkably advanced under many research efforts [5]. Microbatteries, or thin-film batteries, have become commercially available with a rapid expanding market. Nevertheless, similar to the features of macro-scaled batteries, the shortages of microbatteries are limited lifetime and low power density, which bring economic and environmental challenges to systems that they power.

Micro-supercapacitors (m-SC) appeared later as another important energy storage unit. Also known as an electrochemical capacitor (EC), a supercapacitor works through the accumulation of the electrostatic charge within an electrochemical double layer at the electrode/electrolyte interface. It could present high specific capacitance, which mainly depends on the high surface area of the electrode materials, or some pseudo-Faradaic charge transfer process. Compared with the batteries, supercapacitors possess inferior energy density but superior power density, i.e., they can be fully charged/discharged in seconds or minutes. Another prominent advantage of supercapacitors is the long cycle life, which is rather comparable with that of the functional devices. For the large-scale application system, supercapacitors are usually used combined with batteries so that both the high energy density of the batteries and the high power density of superapacitors could be utilized to ensure sufficient power supply. Similarly, m-SC is also complementary but indispensable when high power density is required to support the MEMS devices. More importantly, it could even replace the microbatteries when the cycle life of the device is preferred over the energy density for the whole system. As a matter of fact, with the great development of energy harvesters and nanogenerators, i.e., microscaled energy conversion devices that harvest energy from the ambient environment such as solar power, wind, water flow, vibrational energy, and thermal energy from waste heat, m-SC has been much more competitive as an alternative to batteries to play the role of energy storage in the self-powered micro/nanosystems [6]. **Figure 1** illustrates such a sustainable self-powered system, which consists of five different modules, namely energy harvester, energy storage, sensor, data processor/controller, and data transmitter/receiver [7].

M-SC is originally targeted at high power delivery and robust cyclability, and thus, carbon-based electrode materials are first employed to design on-chip electrochemical double-layer capacitors (EDLC), but the capacitance is relatively low [8–10]. The research interest grows quickly since 2010, when several works were reported on the improved design of the carbon electrode materials, especially the carbide-derived carbon film that possesses high volumetric capacitance and compatible with microfabrication [11–13]. Thanks to the fruitful development of the conventional supercapacitors and nanomaterials, m-SCs also received rapid advances when more research groups turn their interest to the on-chip devices, employing various nanomaterials and designing different fabrication protocols even without being limited to conventional MEMS fabrication routes [14–19].

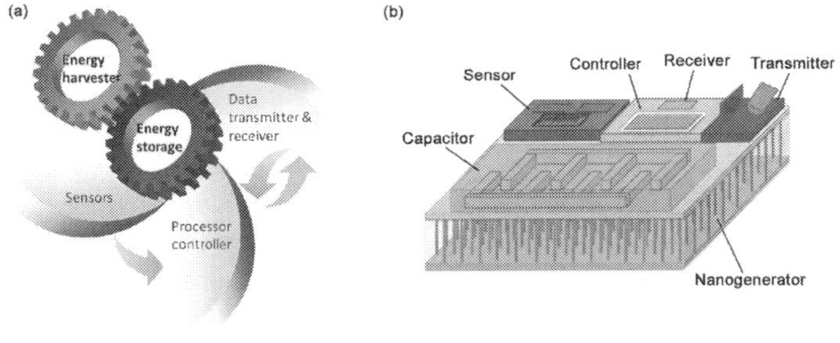

Figure 1. (a) Schematic diagram of the integrated self-powered system showing five modules: energy harvester, energy storage, sensors, data processor & controller, and data transmitter & receiver. (b) Prototype of an integrated self-powered system using a nanogenerator as the energy harvester and a capacitor as the storage unit [7].

After several years of active research, the term of micro-supercapacitor is formally defined and the performance metrics are well recognized [6]. According to the definition given by Beidaghi and Gogotsi, micro-supercapacitors, or electrochemical micro-capacitors, refer to miniaturized supercapacitors that are designed and fabricated to serve as power sources or energy storage units in microelectronic devices. Due to the purpose of being specifically assembled to microelectronic devices, there comes the confinement in the fabrication methods that should be compatible with the current techniques in the semiconductor industry. Hence, the general appearance of m-SCs is a device taking up a footprint area in the millimeter or centimeter scale and a thickness of less than 10 μm. The configurations include sandwiched assemblies consisting of thin-film electrodes, planar arrays of microelectrodes like interdigital electrodes, and three-dimensional (3D) architectures of nanoscale building blocks [6, 12]. The former two configurations are commonly adopted in the present laboratory prototypes, while the latter one, 3D architecture, is a composed idea of the next generation whose realization still requires innovations.

2. Performance metrics for micro-supercapacitors

As perceived to power the microelectronic devices in a self-powered system, sufficient and energy have to be delivered by the m-SCs. The specific requirement of the power depends on the functional devices varying from environmental sensors to personal electronics, which may be in the range of 1–100 μW. Meanwhile, the duration of the power delivery is also required, meaning certain energy has to be supplied. Thus, power and energy together determine the suitability of an m-SC to power a microelectronic system. It should be mentioned that there is another direction emerging for the application of m-SCs other than micropower units, namely replacing electrolytic capacitors in electronic circuits such as alternating current line filtering.

In order to utilize the advantage of miniaturized size of m-SCs to replace the conventional bulky electrolytic capacitors, the m-SCs have to response fast with a relatively large capacitance, i.e., ideal capacitive behavior with a small resistor-capacitor (RC) time constant (e.g., <8.3 ms for ac line filtering). In such case, the capacitance is a critical parameter for the evaluation, and it is also important whether the capacitance is well kept under faster charge/ discharge conditions, which may be termed as rate capability, a terminology from the battery field. In addition, good reversibility (usually assessed by the Coulombic efficiency) and long cycle life should persist for the m-SCs.

As a matter of fact, assuming that an m-SC with a constant capacitance of C is charged from 0 V to an ultimate voltage of U in a duration of t, the stored energy E is calculated by $E = 1/2\ CU^2$. And the power P is calculated by $P = E/t$, which is further written as $P = 1/2\ IU$, if the charge current is I. Therefore, the energy depends on the capacitance and the voltage. Meanwhile, the power depends on the voltage and the working current that is chosen for the operation of the device. However, the current is not chosen arbitrarily, as the capacitance usually decreases with the increase of current. In other words, the energy shows a declining relation with the power, which is usually described as a Ragone plot. In a word, the performance of an m-SC device could be represented by a Ragone plot, or alternatively, it could be described more intrinsically by the voltage and the capacitance versus the working current. Since the footprint area and the occupied volume are limited for m-SCs when integrated into the system, normalized parameters, i.e., the areal or volumetric energy and power densities and capacitance, are the most important in evaluating the m-SCs. The device performance is determined by the intrinsic properties of the electrode materials, the electrolyte, and the device architecture, wherein the electrode materials play the most critical role and attract abundant studies. In order to compare the properties of different electrode materials effectively, the volumetric energy and power densities as well as the volumetric capacitance should be assessed. This is different from the performance metrics for conventional macro-supercapacitors, where gravimetric capacitance, energy, and power densities are emphasized. Due to the microscaled size, the weight of the electrode materials becomes almost negligible, while the volumetric parameters are the most important concern in developing practical devices. On the other hand, mass density is not a limiting factor, which remarkably expands the choices of the electrode materials for novel m-SCs.

3. Electrode materials for micro-supercapacitors

Selection of suitable electrode materials is the kernel in the design of m-SC for certain application. According to the electrochemical energy storage mechanism, the electrode materials currently studied could be divided into three types: carbon-based materials, pseudocapacitive oxides, and conducting polymers. The main purpose of this review is to give a clear comparison among the intrinsic electrochemical properties of the state-of-art electrode materials for m-SC, expecting to open up helpful strategies on developing advanced m-SCs in future.

http://dx.doi.org/10.5772/64986

3.1. Carbon-based materials

Carbon is a typical electrode material for EDLC, which possesses good reversibility and cyclability but limited specific capacitance. For conventional supercapacitors, carbon materials are usually fabricated as powders with a loose structure to facilitate the ions diffusion, so as to acquire a high gravimetric capacitance and fast response rate, but the volumetric capacitance is sacrificed. In addition, it is difficult to prepare uniform and qualified thin films from powders, which hinders the direct application of carbonaceous powders in the microelectrodes. Aiming at carbon-based m-SCs, Chimiola et al. tackle the problem by embracing the technological hurdles [12]. They found the carbide-derived carbon (CDC) films produced by selective etching from metal carbides exhibiting an unprecedentedly high volumetric capacity, holding the promise of developing advanced m-SCs. As a first step, they demonstrated that a thin CDC film on a bulk TiC ceramic plate with strong interface adherence was fabricated by direct chlorination at elevated temperatures. Ti is extracted from TiC as $TiClO_4$, forming a porous carbon film, while the conductive TiC plays as both the substrate and the current collector. Both the porosity of the CDC film and the etching speed are closely related to the chlorination temperature [20]. After 15 s etching at 500°C, a CDC film of 2 μm thick, with a pore size of about 0.7 nm, was synthesized. Its volumetric capacitance reaches 180 and 160 F/cm^3 in an acetonitrile solution dissolving 1 M tetraethylammonium tetrafluoroborate ($TEABF_4$) and an aqueous solution of 1 M H_2SO_4, respectively, which are contributed by almost pure EDLC. They proposed the fabrication processes for m-SCs based on CDC films, which is schematically shown in **Figure 2**. The deposition of the precursor carbides and gold collectors could be conducted by well-known chemical and physical vapor depositions (CDC and PVD). The chlorination and the plasma etching of the photolithography are well-established techniques. Thus, the good compatibility with the semiconductor industry highlighted the promise of CDC-based m-SC devices.

Heon et al. [21] continued the exploration of the electrochemical property of the CDC films, aiming at the CDC-based m-SCs. The synthesis of uniform and adherent porous CDC films on various substrates by reactive DC magnetron sputtering and chlorination was realized, and the high volumetric capacitance of ~180 F/cm^3 in 1.5 M TEABF4/acetronitrile electrolyte was achieved. Later, the on-chip m-SC from CDC films was fabricated and tested by Huang et al. [22]. The preparation process is generally similar to that proposed in **Figure 2a–d** except that Photolithography step is applied on the TiC film to produce interdigitated electrodes before the chlorination and the current collectors deposited on the CDC electrodes are Ti/Au layers instead of the single Au layer. The active material, i.e., TiC-CDC film, was 1.6 μm thick. The device was dipped into 1 M NEt_4BF_4 in propylene carbonate (PC) electrolyte in glove box under Ar atmosphere for electrochemical tests, exhibiting the EDLC behavior over a potential window of 2 V with an areal capacitance of 1.5 mF/cm^2, a maximum energy density of 3.0 mJ/cm^2, and a maximum power density of 84 mW/cm^2. At this point, the EDLC property of TiC-CDC film and the feasibility of manufacturing on-chip m-SCs that are to be integrated into MEMS and electronics have been fully demonstrated, which should be recognized as a representative research on the development of carbon-based m-SC device. Nevertheless, the energy and power performance of the TiC-CDC m-SC are within the range of values reported for other carbon-based micro-supercapacitors, in spite of the high volumetric capacitance of the TiC-CDC active material [10, 22–24]. This is fundamentally attributed to the intrinsic limitation of the capacitance inhibited from the EDLC behavior, which only involves the variation of the ion concentrations in the electric double layer, and the theoretical capacitance is 16–50 μF/cm^2.

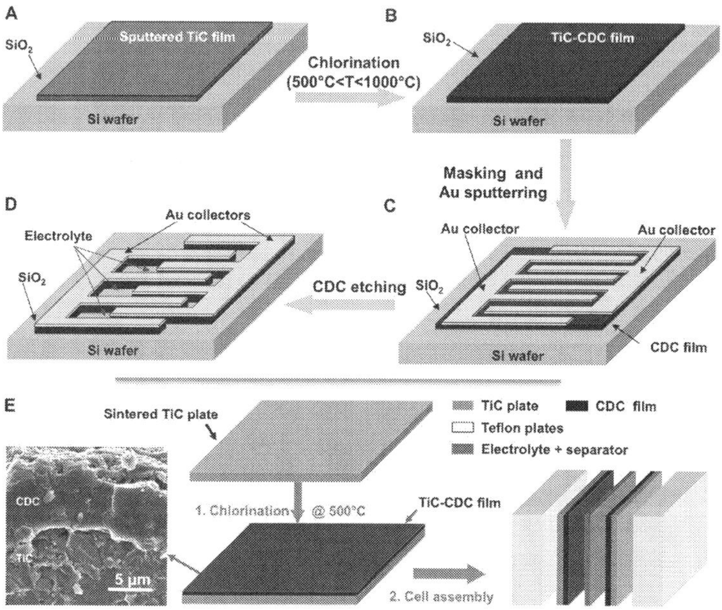

Figure 2. (A-D) Schematic of the fabrication of an on-chip m-SC based on the CDC film process, using standard photolithography. (E) Schematic of CDC synthesis and a sandwiched device for electrochemical test. The SEM micrograph shows the interface of CDC/TiC. Reproduced with permission from ref. 12. Copyright (2010) American Association for the Advancement of Science.

To acquire high surface area within the limited electrode volume is still the most effective way to obtain high capacitance for the carbonaceous m-SCs. In this case, activated carbon is still a good choice. However, many well-developed fabrication routes of activated carbon powders are not easily applicable for synthesis of thin activated carbon films (ACFs) which is necessary for m-SCs. The challenges include the formation of cracks in the carbon film due to the shrink of polymers during heating and carbonization, the weak interface between the polymer and the substrate resulted from the large stress produced under the harsh synthesis and cooling conditions, and the possible damage to the brittle film in the photolithographical process. Wei et al reported an effective way to minimize some of the interface stresses in order to fabricate ACFs, namely catalyst-assisted low-temperature carbonization of an organic compound solution [25]. The sucrose and H_2SO_4 (as a catalyst) aqueous solution was spin-coated onto a silicon wafer, dried at room temperature, carbonized, annealed at 700°C in vacuum to remove decomposition products of the carbohydrate and catalyst residues, and activated by annealing at 900°C in CO_2 to induce open porosity within carbon. ACFs of 1–2 μm thick were thus produced free of microcracks, while the interface adhesion to the SiO_2/Si wafer could be reinforced by further annealing at 1100°C in Ar, which could survive lithographical patterning as evidenced experimentally. The electrochemical property of such ACF film electrodes was tested in a symmetric sandwich-type configuration with 1 M H_2SO_4 as the electrolyte solution. Exhibiting typical EDLC property by the rectangular cyclic voltammetry (CV) curves, an extremely high volumetric capacitance of 390 F/cm^3 was obtained under the slow scan rate of 1 mV/s, which is the highest value reported for carbon film electrode materials at present (see **Figure 3**). Moreover, **Figure 3** shows clearly that the performance was strongly affected by the activation time. As the film becomes more porous after longer activation time, the capacitance increases and the rate capability improves as well, which should be due to easier accessibility of ions into the films. It means that the volumetric metrics of the carbonaceous electrode materials could be greatly enhanced when the materials structure is carefully optimized with proper fabrication techniques. But it is also worth noting that the fabrication process is still quite harsh as several times of high temperature annealing are required, which probably causes difficulties for the practical manufacturing of the devices integrated on the chips, although the realization of uniform and robust ACF films represents a great progress itself.

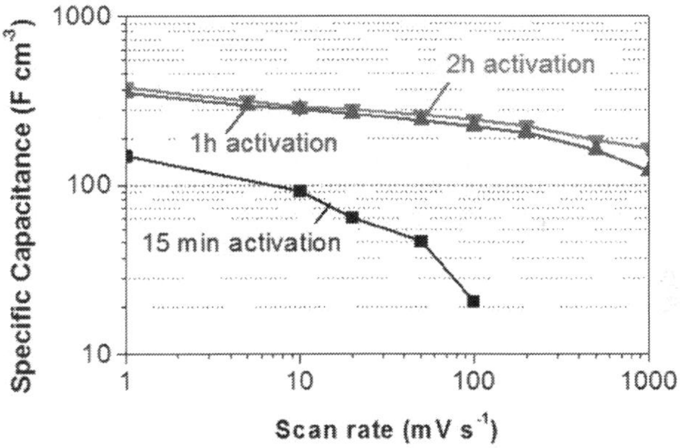

Figure 3. Rate performance of symmetric ACF electrode cells in 1 MH_2SO_4 electrolyte: volumetric capacitance of ACFs as a function of CV scan rate. Reproduced with permission from ref. 25. Copyright (2013) American Chemical Society.

As a matter of fact, there have also been great advances in the research of carbon-based electrode materials for macro-scaled supercapacitors that pursue higher volumetric property in recent years. For example, liquid-mediated dense graphenes [26] and nitrogen-doped mesoporous carbon [27] were reported to have unprecedentedly high volumetric capacitance that is comparable with pseudocapacitive materials. Pseudocapacitance has been induced on these modified carbon materials actually [28]. Yang et al. tackle the problem by considering the paradox between gravimetric capacitance C_{wt} and packing density of the carbon ρ [26]. The specific volumetric capacitance C_{vol} is simply calculated by $C_{vol} = C_{wt} \times \rho$, while C_{wt} is always compromised with the increase of ρ. They addressed this challenge with liquid electrolyte-mediated chemically converted graphene (EM-CCG) films. They started with the chemically reduced graphene oxide, namely chemically converted graphene (CCG) sheets, which are well dispersed in water and could self-assemble to form an oriented hydrogel film through a directional flow-induced bottom-up assembly process. With the CCG sheets remaining largely separated, a high C_{wt} of over 200 F/g was obtained, while the packing density was only ~0.069 g/cm^3, resulting in a mediocre C_{vol} of ~18 F/cm^3. In order to compress the CCG hydrogel films, they were exchanged with a miscible mixture of volatile and nonvolatile liquids and then subjected to removal of the volatile liquid by vacuum evaporation. As a consequence, the film thickness was reduced, while the sheets remained solvated by the nonvolatile liquid (e.g., sulfuric acid). Through electrochemical tests in 1 M H_2SO_4 electrolyte, they found that the C_{vol} of the EM-CCG films was nearly proportional to ρ, and the highly compact one with ρ of 1.33 g/cm^3 yielded a C_{vol} of 255.5 F/cm^3 at 0.1 A/g, which is much higher than previous porous carbon materials for conventional SCs. Although the adaptable intersheet spacing among the graphene sheets is particularly emphasized in this work to optimize the C_{vol}, other factors such as surface wettability and the pore interconnectivity are also important to realize superior capacitive property for the carbon materials [28]. In other words, both surface chemical property and structural configuration play significant roles in determining the volumetric capacitance of the carbonaceous electrode materials. Lin et al. made a breakthrough in the chemical way, finding that a nitrogen-doped ordered mesoporous few-layer carbon has an extremely high specific capacitance of 855 F/g when tested in 0.5 M H_2SO_4 electrolyte at 1 A/g (comparing with the 200 F/g for the CCG sheets) [27]. The extra capacitance comes from the pseudocapacitance contributed by the doped N in the pyrrolic and pyridine sites incorporating protons. To be assembled into supercapacitors, they studied variation in the volumetric capacitance with the mass loading of this N-doped mesoporous carbon materials, which showed that the highest value of 560 F/cm^3 was reached at the loading of 6 mg/cm^2. Although great advances have been accomplished in the carbonaceous materials for the macro-scaled SCs, they are in the form of either self-standing membranes or powders, difficult to be integrated into the on-chip m-SCs.

Besides the attempts to enhance the volumetric capacitance for energy storage through a proper film fabrication process, there is another important direction for carbon-based m-SC, i.e., to utilize its fast response rate to replace electrolytic capacitors by making open-structured devices [29, 30]. Pech et al. produced m-SCs through electrophoretic deposition of a several-micrometre-thick layer of nanostructured carbon onions (OLC) with diameters of 6–7 nm onto the interdigital Au current collectors patterned on silicon wafers. The OLC particles were prepared by annealing nanodiamond powder at 1800°C. A stable capacitive behavior was obtained for the microdevice over a 3 V potential window in a 1 M solution of tetraethylammonium tetrafluoroborate in PC, with a linear dependence of the discharge current on the scan rate and low resistive contributions up to 100 V/s, which is about three orders of magnitude higher than conventional SCs. Such a microdevice preserves an areal capacitance of 0.9 mF/cm^2 at 100 V/s, which is comparable to values usually reported at much lower scan rates (1–100 mV/s) for microscaled EDLC devices (0.4–2 mF/cm^2) [9, 11, 31]. The most appealing feature of such OLC m-SC is the extremely small characteristic relaxation time constant τ_0 (the minimum time needed to discharge all the energy from the device with an efficiency of greater than 50%), which is only 26 ms, much lower than that of the AC-based microdevice (τ_0 = 700 ms) or OLC-based macroscopic devices (τ_0 > 1 s) [32]. **Figure 4** shows the Ragone plot of several typical energy storage devices designed for power microelectronics applications, including a 500-μAh thin-film lithium battery, a 25-mF supercapacitor, and an electrolytic capacitor of the same absolute capacitance, as well as the m-SCs composed of AC, OCL and graphene-based materials. It could be seen that the power density of the OLC-based m-SC has reached that of the electrolytic capacitors, but the energy density is more than one order of magnitude higher than that of latter.

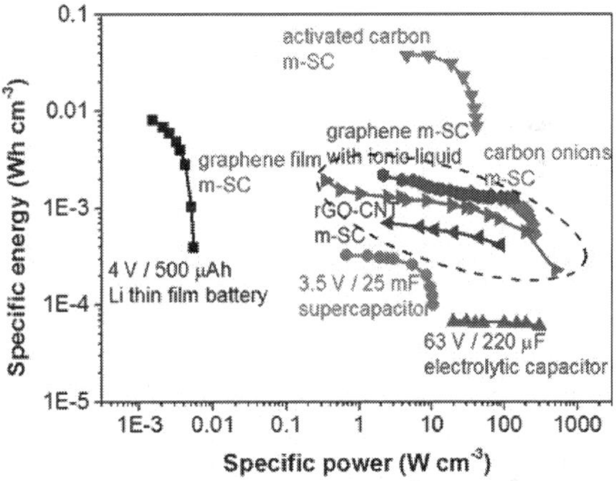

Figure 4. A Ragone plot showing the relationship between the volumetric energy density and power density of typical electrolytic capacitors, supercapacitors, batteries, and the m-SCs with AC and OLC electrode [29], as well as the m-SCs with various graphene films [17, 33] and graphene-CNT (rGO-CNT) composite electrode materials [34]. The dashed ellipsoid generally describes the best high power performance currently achieved by these state-of-art m-SCs assembled with conductive carbon electrode materials.

Thereafter, several research papers reported high-power m-SCs fabricated from graphene, whose performance reaches a similar level with that of the OLC-based device (see **Figure 4**). For example, interdigitated graphene m-SCs were produced through laser burning along designed patterns on a graphene oxide (GO) film supported on a PET sheet which was inserted into a LightScribe DVD drive [17]. Due to the photo-thermal effect under laser radiation, the exposed GO was converted into graphene, constructing the positive and negative electrodes, while the unexposed GO remained insulating and served as a separator. A hydrogel-polymer electrolyte, poly(vinyl alcohol) (PVA)-H_2SO_4, was then drop-cast on the patterned area to create a planar m-SC. Such a device using reduced GO (rGO) as electrode materials exhibits an areal capacitance of 2.32 mF/cm^2 and a volumetric capacitance of 3.05 F/cm^3, with a characteristic relaxation time of only 19 ms and a high power density of nearly 200 W/cm^3. There is another planar device using rGO and carbon nanotube (CNT) composites as the electrode material, which is prepared by combining electrostatic spray deposition (ESD) and photolithography lift-off methods [33]. The m-SC delivers an areal capacitance of 6.1 mF/cm^2 at 10 mV/s, and a value of 2.8 mF/cm^2 is still preserved at 50 V/s, corresponding to 3.1 F/cm^3. Its characteristic time constant is only 4.8 ms. An even faster m-SC device is made from graphene films of only 6–100 nm thick, whose maximum capacitance is 0.807 mF/cm^2 and 17.9 F/cm^3 (specific values of 0.323 mF/cm^2 and 71.6 F/cm^3 for the electrode material), with the maximum power density reaching 495 W/cm^3, the maximum energy density 2.5 mWh/cm^3, and the characteristic time constant as short as 0.28 ms [34]. It is concluded from these researches that the electronic conductivity of the electrodes has to be enhanced in order to acquire fast response performance. The most straightforward route is to reduce micropores within the electrodes and enlarge the open area in direct contact with the electrolyte, as the outmost surface is the most easily accessible with the ions for charge/discharge processes. However, such a design is usually at the cost of volumetric capacitance, and the film thickness should be thin as well, which thus limits the areal capacitance of the device.

3.2. Pseudocapacitive oxides

For the pseudocapacitive materials, Faradaic charge transfer occurs on the electrode/electrolyte interface during charge/discharge processes, giving much higher areal capacitance than EDLC does. Many transition metal oxides exhibit pseudocapacitive behavior in certain aqueous electrolyte. RuO_2 is the first discovered pseudocapacitive oxide and still the most ideal candidate [35, 36]. Within the potential range of 0–1.4 V vs. SHE, RuO_2 continuously changes its valence from Ru^{2+} to Ru^{4+}, following the reaction of $RuO_2 + xH^+ + xe^- \leftrightarrow RuO_{2-x}(OH)_x$, where $0 \leq x \leq 2$ [37]. The process undergoes through both electron transfer and proton incorporation in RuO_2 particles. Because of the good electronic conductivity and proton conductivity for hydrated RuO_2, fast and reversible charge/discharge pseudocapacitance according to adsorption isotherm model [38] is observed, with a specific capacitance value over 600 F/g. The practical capacitance is closely related to the crystallinity of the material. For crystalline RuO_2, protons only adsorb on the surface instead of entering the grains, which provides a capacitance per real surface area of 339–490 μF/cm^2 and an overall specific capacitance of about 380 F/g [37]. For amorphous $RuO_2{\cdot}xH_2O$, protons could easily transport inside the domains, thus presenting a much higher specific capacitance.

There have been researches on RuO_2 thin-film electrodes ever since 1990s. Jow and Zheng coated an amorphous RuO_2 film onto Ti substrate through sol-gel method. In spite of inferior uniformity and many cracks, the film still shows a capacitance of 40 mF/cm^2 as tested at 50 mV/s in 0.5 M H_2SO_4, which decreases by only 10% at 500 mV/s [37]. Zheng et al. further employed pulsed laser deposition (PLD) to prepare the RuO_2 films [37]. The amorphous film deposited at room temperature possessed the highest capacitance (6.3 mF/cm^2). As the deposition temperature increased to above 200°C, the capacitance reduced to only 0.3 mF/cm^2. The diffusion length of protons in amorphous RuO_2 film was estimated to be about 5.8 nm, while it reached >11.1 nm in $RuO_2{\cdot}xH_2O$. Assuming the diffusion coefficient to be $>10^{-8}$ cm^2/s, the proton in and out diffusion from RuO_2 film is in the order of 10 μs for a diffusion length less than 11.1 nm. This explains the good rate capability of the RuO_2 electrode (i.e., it could be charged/discharged at a rate of over 500 mV/s without loss of the capacitance). Thus, for the RuO_2 electrode, the charge/discharge rate is mainly limited by the electric resistance and the proton transport in the electrolyte, rather than the proton diffusion inside the RuO_2 material.

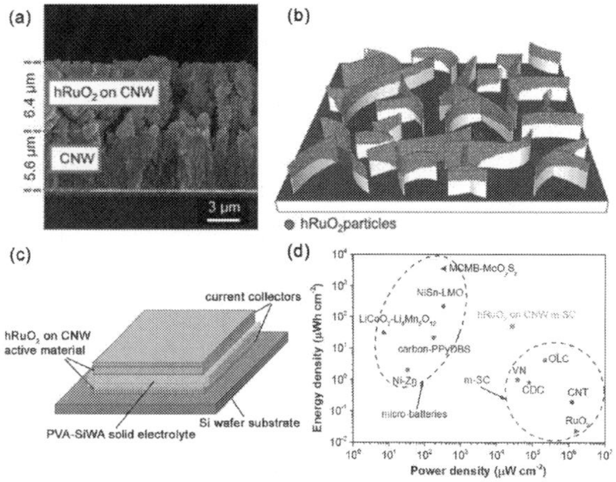

Figure 5. (a) Cross-section SEM image of a CNW film (bottom) electrodeposited by $hRuO_2$ on its top. (b) Schematic of the vertically aligned CNW decorated with $hRuO_2$ particles. (c) Schematic diagram of on-chip m-SC with 2D architecture. (d) A Ragone plot showing the energy and power density of the CNW/$hRuO_2$-based m-SC, compared with other advanced m-SCs and microbatteries. Reproduced with permission from ref. 39. Copyright (2014) Elsevier Ltd.

The development of RuO_2-based m-SCs emerged in recent years. Liu et al. [36] fabricated the planar device through depositing RuO_2 nanorods onto the patterned stack layer of Ru/Au/Ti/SiO_2 on silicon wafer, which was subjected to electrodeposition for another layer of hydrous RuO_2 on top. It worked well in 0.5 M H_2SO_4 electrolyte, providing a capacitance of 21.4 mF/cm^2 at 50 mV/s and 14.9 mF/cm^2 at 500 mV/s. Makino et al. [18] reported the fabrication of an m-SC with ordered mesoporous RuO_x as the electrode material, which was produced by controlled electro-deposition using a lyotropic liquid crystal template method and subsequent electro-oxidation on an interdigital electrode array. The device exhibited good capacitive property with maximum capacitance of 12.6 mF/cm^2 and maximum energy of 1.49 μWh/cm^2 at the slowest discharge rate of 0.38 mA/cm^2 and maximum power delivery of 750 μW/cm^2 at 2.88 mA/cm^2. More recently, a new m-SC device based on hydrous RuO_2/carbon nanowalls hierarchical structured composite electrode was proposed by Dinescu et al., which showed an exceptionally high capacitance [39]. Carbon nanowalls (CNW), or vertically oriented graphene sheets, is a good EDLC material with a large surface area, good electronic conductivity, and excellent chemical stability, while RuO_2 is an ideal pseudocapacitive materials with a high specific capacitance. A silicon wafer coated with an insulating Si_3N_4 layer was first deposited with a 40 nm Cr/200 nm Pt layer by evaporation as the current collector and subjected to the CNW layers growth by PECVD at 700°C, and then, electrodeposition of hydrous RuO_2

($hRuO_2$) onto the CNW was carried out afterwards, after which the samples were annealed in air at 150 °C. The pristine CNW layer is 12 μm thick, with a capacitance of 5.7 mF/cm^2 in 0.5 M H_2SO_4 electrolyte, close to the values of other carbonaceous electrodes. When about half of the CNW layer was decorated with $hRuO_2$ (see **Figure 5a, b**), the hybrid electrode exhibited an extremely high capacitance of 1094 mF/cm^2 at 2 mV/s, which is three orders of magnitude higher than that of the state-of-the-art graphene-based m-SCs [34], and also far larger than most other advanced m-SC electrodes [22, 25, 33, 40]. An all-solid-state m-SC in a stack configuration was realized with a solid-polymer electrolyte sandwiched between two CNW/$hRuO_2$ electrodes (see **Figure 5c**), delivering an energy density of 49 μWh/cm^2, i.e., 20 mWh/cm^3. Such a value is even comparable to the state-of-the-art lithium ion microbatteries [41–43], but its power density and cycle life (more than 90% is retained after 2000 cycles) are much higher than that of the latter, which is shown in **Figure 5d**.

In spite of the ideal pseudocapacitve property, RuO_2 is too expensive for large-scale application. Cheaper oxides have been widely researched, such as MnO_2 and NiO, wherein MnO_2 attracts the most attention [44, 45]. MnO_2 works in neutral aqueous solutions, with the potential window within 0.8 V and above 0 V vs. Ag/AgCl, and thus is suitable to serve as a positive electrode in asymmetric devices. Its working mechanism is the surface adsorption/desorption of the electrolytic cations and protons in the solution, which is described as the reaction $MnO_2 + xC^+ + yH^+ + (x+y)e^- \leftrightarrow MnOOC_xH_y$. [46, 47]. It is obvious from the CV curve on **Figure 6** that the charge/discharge behavior of MnO_2 is similar to that of EDLC. Besides, the abundant resource and the safe working condition of neutral solutions also boost the wide research on MnO_2, although the specific capacitance of MnO_2 powders or micrometer-thick films is only 150 F/g.

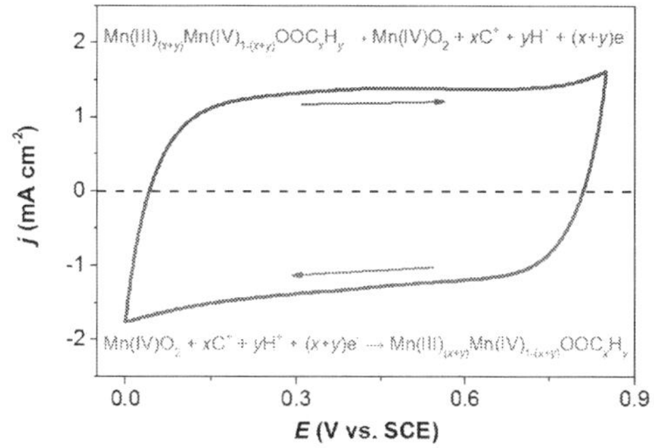

Figure 6. The CV curve of MnO_2 electrode tested in 2 M Li_2SO_4 aqueous electrolyte solution..

The intrinsic electrochemical properties of MnO_2 films produced by different preparation conditions and of different morphologies and structures have been explored [48–51]. For example, the MnO_2 film deposited by cathodic electrodeposition is poorly crystalline and porous, whose specific capacitance strongly decreases with the scan rate and the film thickness [51]. The specific capacitance of a film bearing a deposited mass of 45 μg/cm^2 is tested to be 353 F/g (15.9 mF/cm^2) at 2 mV/s and 135 F/g (6.1 mF/cm^2) at 100 mV/s. For films annealed under 200 °C, whose porosity is reduced and crystallinity increased, the maximum specific capacitance is decreased, while rate capability is improved more or less. PLD has also been utilized to prepare manganese oxide films, with amorphous MnO_x, crystalline Mn_2O_3, and Mn_3O_4 to be produced under different deposition temperatures and the partial pressure of oxygen [50]. The crystalline Mn_2O_3 film possesses the highest specific capacitance, 210 F/g at 1 mV/s for a film of 120 nm, while the Mn_3O_4 film has the lowest value.

Because of the insulating property of MnO_2 and the increased difficulty for ions to access into a thicker film, the rate capability of MnO_2 film is always unsatisfying. Research efforts have been devoted to enhance the electronic conductivity so as to improve the rate capability. Si et al. [52] deposited the MnO_x/Au multilayer film, which showed a capacitance of 32.8 F/cm^3, higher than that of the pure MnO_x film electrode (19.9 F/cm^3). Two kinds of interdigital m-SCs were prepared using these two films. The device of MnO_x/Au multilayer possessed a smaller equivalent series resistance (ESR) and a shorter characteristic relaxation time (5 ms). Doping is another way to adjust the electrochemical property. By doping Mo into the electrodeposited manganese oxide, a film of $MnMo^{6+}{}_{0.18}O_{1.18}(OH)_{0.59}(H_2O)_{0.25}$ was obtained, whose specific capacitance (190.9 F/g and 18.5 mF/cm^2 at 5 mV/s), cycling reversibility, and rate capability were all improved as compared with the undoped film [53]. It was found that the electronic resistivity of a ~0.5-μm-thick film was reduced from 5.0×10^4 to 3.0×10^1 Ω cm after being doped with Mo, which suggested that the enhancement of the electrochemical property is mainly attributed to the increase in electronic conductivity. Similar effect was also discovered in doping Co into the electrodeposited manganese oxide film and in doping V into the PLD-deposited amorphous manganese oxide film [54, 55].

Other pseudocapacitve oxides include Co_3O_4, NiO, $NiCo_2O_4$, and so on [56–58]. They could reversibly form hydroxides in alkaline solutions, the process of which could be represented as $M_3O_4 + OH^- + H_2O \leftrightarrow 2MOOH + e^-$, wherein M refers to elements such as Ni and Co. Their specific capacitances are reported to be very high (382–1400 F/g); however, phase transformations are always involved during the reaction processes, and the charge/discharge behaviors are more like that of batteries, for example, with potential plateaus, short potential windows (0.4–0.5 V), limited rate capabilities [56].

Overall, pseudocapacitive oxides could exhibit high specific capacitance, but poor rate capability and cycling stability hinder their application, which is mainly attributed to their insulating property. For the film electrodes, when thickness increases, the areal capacitance rarely scales up as expected. To disperse the active materials onto a 3D current collector is a common way to acquire certain areal capacitance with acceptable rate capability, which relies on the innovation of nanofabrication techniques. Nevertheless, the intrinsic relationship between the comprehensive electrochemical property and the microstructure of the oxides is still worth studying, which may help to optimize the intrinsic electrochemical performance of the film electrodes, contributing to the development of m-SCs compatible with conventional microelectronics manufacturing techniques.

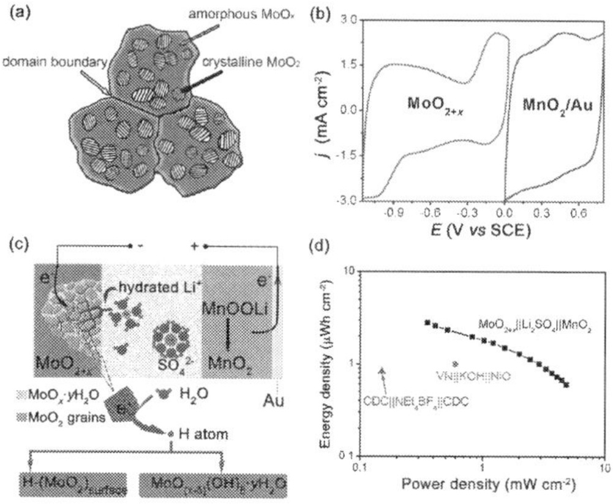

Figure 7. (a) Schematic of the multi-phased microstructure of the MoO_x film deposited at 150°C by magnetron sputtering. (b) CV curves of MoO_{2+x} and MnO_2 films in 2 M Li_2SO_4 electrolyte, at 50 mV/s. (c) Illustrative diagram for the working process and (d) a Ragone plot showing energy density vs. power density of MoO_{2+x}(−)//2 M Li_2SO_4//MnO_2(+) microdevice. The performances of representative m-SCs of EDLC type (symmetric device based on CDC electrode [22]) and other pseudocapacitive type (asymmetric device based on VN and NiO electrodes [61]) are also plotted for reference [59, 60] Reproduced with permission from ref. 60. Copyright (2014) Elsevier Ltd.

Our group has systematically studied the fabrication and electrochemical property of molybdenum oxide thin film, which has various chemical valences with quite different properties [59]. The composition, microstructure, and morphology were controlled to enhance the electrochemical performance of the molybdenum oxide film, and its potential to be applied as a superior electrode material in m-SC is evaluated. We fabricated electronically conductive MoO_{2+x} films via RF magnetron sputtering from a MoO_3 target. Multi-valence composition and mixed-phased microstructure, i.e., coexistence of MoO_2 nanocrystals and amorphous MoO_x ($2 < x \leq 3$), were acquired in these films (see **Figure 7a**), which exhibit excellent pseudocapacitance in Li_2SO_4 electrolyte [59]. The MoO_x ($x \approx 2.3$) film deposited at 150°C presented an areal capacitance of 31 mF/cm^2 at 5 mV/s, corresponding to a volumetric value of 392 F/cm^3, superior to most of the advanced m-SC electrode materials previously discussed. Forty-seven percent of the capacitance was retained when the scan rate increases from 20 to 500 mV/s, meaning good rate capability. The cycling stability is excellent as well, with 100% preserved after 5000 cycles. The multi-phased microstructure of such MoO_x films is quite interesting, as it intrinsically endows the material with superior electrochemical property. The pseudocapacitance originates from the cation (H^+ and Li^+) insertion/extrusion in the amorphous MoO_x, in which H^+ is more active. The MoO_2 grains could also catalyze the decomposition of water combined on the surface, producing H atoms that could be reversibly stored in amorphous MoO_x and thus promote the pseudocapacitive process. Furthermore, the crystalline MoO_2 also improves the electronic conductivity and maintains a stable structure of the film [60]. Another interesting feature about MoO_x film is its relatively negative potential window, i.e., between −1.1 and 0 V vs. SCE, which makes it a proper anode material relative to other pseudocapacitive oxides. Thus, an asymmetric microdevice of MoO_{2+x}(−)//2M Li_2SO_4//MnO_2(+) is successfully fabricated.

http://dx.doi.org/10.5772/64986

It could be seen from **Figure 7b** that the working potential windows of the two electrodes well complement to each other. The working mechanism is described in the schematic of **Figure 7c**. A high energy density of 2.8 μWh/cm^2 at a real power density of 0.35 mW/cm^2 was obtained from this device, combined with good stability (no capacitance loss for 10,000 cycles). The Ragone plot of **Figure 7d** shows that the performance of this asymmetric device is much better than other typical state-of-art m-SCs.

Microelectrode[a]	Set up	Thickness	C_V (F cm^{-3})[b]	τ_c (ms)	Cyclability	References
Activated carbon	Two-electrode assembly in 1 M H_2SO_4	1.5 μm	390 (1 mV s^{-1})	1111	95% after 10,000 cycles at 10 A g^{-1}	[25]
TiC-CDC nanofelts	Filled into a micro cavity electrode tested in 1 M H_2SO_4	125 μm in diameter, 35 μm in depth	12 (10 mV s^{-1})	379	Stable over 10,000 cycles	[64]
Polyaniline nanowire arrays	Interdigital micro device with H_2SO_4 -PVA gel	400 nm	588 (0.1 mA cm^{-2})	–	96% after 1000 cycles	[16]
rGO-CNT composite	Interdigital micro device with 3 M KCl	6 μm	37.5 (10 mV s^{-1})	4.8	Slight decline over 1000 cycles	[33]
rGO	Interdigital micro device with H_3PO_4 /PVA gel	25 nm	359 (1 A g^{-1})	–	90% after 1000 cycles	[65]
Photoresist-derived porous carbon	Three-electrode assembly in 3.5 M KCl	1 μm	35 (10 mV s^{-1})	–	No loss after 10,000 cycles at 100 mV s^{-1}	[66]
TiC-CDC	Three-electrode assembly in 1 M H_2SO_4 or 1 M $TEABF_4$	~2 μm	160 (in H_2SO_4); 180 (in $TEABF_4$)	–	–	[12]
MnO_x/Au multilayers	Interdigital micro device with H_2SO_4-PVA gel or 1 M Li_2SO_4	50 nm	78.6 (10 mV s^{-1}, H_2SO_4-PVA gel); ~400 (10 mV s^{-1}, Li_2SO_4)	5	74.1% retained after 15,000 cycles at 1 V s^{-1}	[52]
rGO	Interdigital microdevice with H_2SO_4/PVA gel	15 nm	71.6 (10 mV s^{-1})	0.28	98.3% retained after 10^5 cycles at 50 V s^{-1}	[34]
CNT film	Three-electrode assembly in 1 M H_2SO_4	30–250 nm	132 ± 8 (50 mV s^{-1})	–	–	[67]
Laser-scribed graphene	Interdigital microdevice with H_2SO_4-PVA gel	7.6 μm	~12.2 (10 mV s^{-1})	19	94% retained after 10,000 cycles	[17]
Onion-like carbon	Interdigital microdevice in 1 M Et_4NBF_4/PC	7 μm	~9.7 (1 V s^{-1})	26	Stable over 10,000 cycles at 10 V s^{-1}	[29]
Activated carbon	Interdigital microdevice in 1 M Et_4NBF_4/PC	5 μm	~92.8 (500 mV s^{-1})	700	–	[29]
Electrodeposited MoO_x	Three-electrode assembly in 2 M Li_2SO_4	120 nm	1162 (5 mV s^{-1})	1381	88% retained after 4000 cycles at 100 mV s^{-1}	[62]
Electrodeposited MoO_x after annealing at 350°C for 1.5 h	Three-electrode assembly in 2 M Li_2SO_4	63 nm	700 (5 mV s^{-1})	11	99% retained after 4000 cycles at 100 mV s^{-1}	[62]

[a]Carbide-derived carbon (CDC), reduced graphene oxide (rGO), carbon nanotubes (CNT).
[b]Please note that C_V here corresponds to the electrodes, and some of the values are read from the reported plots, or estimated from the cell data.

Table 1. Electrochemical properties of typical m-SC microelectrodes in literature [62].

The pseudocapacitive MoO_x film was further prepared by electrodeposition, with its electrochemical property adjusted by annealing under different conditions [62]. Optimal experimental parameters were determined to fabricate the film containing MoO_2 nanocrystallites and amorphous MoO_x. A film of 63 nm thick exhibited a high volumetric capacitance of 700 F/cm^3, good rate capability with a relaxation time constant of 11 ms, and excellent cycling stability of 99% capacitance retention after 4000 cycles. The performance is superior to other typical microelectrodes for m-SCs, the comparison of which is listed in **Table 1**. Furthermore, a 3D microelectrode was developed by electrodepositing MoO_x on a Ti nanorod array prepared by oblique angle deposition [63]. An areal capacitance of 27 mF/cm^2, corresponding to a high volumetric capacitance of 643 F/cm^3, was obtained as well as satisfying cycling stability, which is rather attractive compared to other 3D microelectrodes. And post-annealing in reductive atmosphere improved its rate capability and response speed. Thus, the further improvement in electrochemical property of MoO_x electrode by architecture design of employing current collectors with large specific area promotes its practical application in m-SCs.

3.3. Conducting polymers

Conducting polymers are another group of pseudocapacitive materials, which work through the fast redox reaction of ion doping [68, 69]. Currently studied conducting polymers mainly include P-type doping materials such as polypyrrole (PPy) and polyaniline (PANI), and N-type doping materials such as polythiophene (PTH). They are electronically conductive, leading to low ESR, and their capacitance is generally 2–3 times as high as that of activated carbon materials. However, the cycling stability is always unsatisfying due to the large volume expansion and shrink during the charge/discharge processes. The application of conducting polymers in m-SCs is relatively pioneering, i.e., developed in 3D structures [16, 70, 71]. Beidaghi and Wang [71] fabricated such a 3D-structured interdigital m-SC through carbon-microelectrochemical system (C-MEMS) technology. Briefly, the carbonization of patterned photoresist pillars produced microarrays of carbon pillars on the interdigital carbon layer supported on silicon wafer. The carbon arrays served as the C-MEMS current collectors, on which PPy was coated by electrodeposition. The resulted PPy/C-MEMS electrode presented a high areal capacitance of 162 mF/cm^2 (volumetric capacitance estimated to be 11.6 F/cm^3) and a power density of 1.62 mW/cm^2 at 20 mV/s scan rate. Correspondingly, the entire symmetric m-SC device exhibited an average capacitance of 78 mF/cm^2 and a power density of 0.63 mW/cm^2. In spite of the high areal capacitance, the electrodes were not robust for consecutive cycling, as only 56% of the capacitance is retained after 1000 cycles.

4. Summary

Architecture design is especially important for m-SC, which significantly affects the comprehensive device performance. Structure of a separator layer sandwiched by thin-film electrodes is the traditional architecture, but the performance is closely related to the thickness of the electrode layer, which hinders it from scaling up to acquire higher areal energy density [12]. In-plane interdigital architecture is often employed in recent years, especially for fast response devices [13, 17]. Besides, integrated 3D architectures have also been proposed for m-SCs [19]. The microelectrodes with the structure of micro- or nanoarrays as cited above [16, 63, 70, 71] could be considered as ordered 3D structure, which is already realized by microfabrication techniques. However, a "true" 3D device is composed to consist of interpenetrating electrodes that are separated by a very thin layer of electrolyte, which is proposed by Long and coworkers [5]. Although several possible strategies for fabrication of electrochemical energy storage devices with 3D architectures have been proposed [72], there has been no work fully realized the concept of 3D m-SC [6].

Above all, structure control is still the most important factor in exploring the property limits of different kinds of electrode materials. Although excellent efforts have been made on developing carbon film electrode with better volumetric and areal capacitances, the value is still lower than that of pseudocapacitive materials, which is limited by the theoretical limits of EDLC. For the pseudocapacitive materials such as molybdenum oxide, it is possible to optimize the intrinsic electrochemical property by structure design, i.e., to obtain both good rate capability and cyclability as well as keep relatively high capacitance. However, the application of the oxides is seriously limited by the electrolyte condition, which is always a disadvantage compared with carbon materials. Pseudocapacitive carbon materials represent another promising trend to achieve balanced performance. In-plane m-SCs with interdigital architectures have proven its success to fabricate fast response devices with carbon electrodes. Thus, further study on the interplay effect between such architecture and pseudocapacitive materials is in need. Exploring 3D architecture for m-SC is a difficult but attracting challenge.

Acknowledgements

The authors are grateful to the financial support by the National Basic Research Program of China (973 program, Grant No. 2013CB934301), the National Natural Science Foundation of China (Grant No. 51531006 and No. 51572148), the Research Project of Chinese Ministry of Education (Grant No. 113007A), and the Tsinghua University Initiative Scientific Research Program.

Author details

Can Liu[1] and Zhengjun Zhang[2*]

*Address all correspondence to: zjzhang@tsinghua.edu.cn

1 Institute for Catalysis, Hokkaido University, Sapporo, Japan

2 School of Materials Science and Engineering, Key Laboratory of Advanced Materials, Tsinghua University, Beijing, People's Republic of China

References

[1] K.B. Lee, L.W. Lin, Electrolyte-based on-demand and disposable microbattery, J Microelectromech S, 2003;12:840–847. doi:10.1109/Jmems.2003.820272

[2] A.H. Epstein, S.D. Senturia, G. Anathasuresh, A. Ayon, K. Breuer, K.S. Chen, F. Ehrich, G. Gauba, R. Ghodssi, C. Groshenry, S. Jacobson, J. Lang, C.C. Mehra, J.M. Miranda, S. Nagle, D. Orr, E. Piekos, M. Schmidt, G. Shirley, S. Spearing, C. Tan, Y.S. Tzeng, I. Waitz, Power MEMS and microengines, in: 1997 International Conference on Solid State Sensors and Actuators, 1997. TRANSDUCERS '97 Chicago, vol. 2, 1997, pp. 753–756 vol. 752.

[3] A.H. Epstein, S.D. Senturia, Macro power from micro machinery, Science, 1997;276:1211–1211. doi:10.1126/science.276.5316.1211

[4] A. Mehra, Z. Xin, A.A. Ayon, I.A. Waitz, M.A. Schmidt, C.M. Spadaccini, A six-wafer combustion system for a silicon micro gas turbine engine, J Microelectromech S, 2000;9:517–527. doi:10.1109/84.896774

[5] J.W. Long, B. Dunn, D.R. Rolison, H.S. White, Three-dimensional battery architectures, Chem Rev, 2004;104:4463–4492. doi:10.1021/cr020740l

[6] M. Beidaghi, Y. Gogotsi, Capacitive energy storage in micro-scale devices: recent advances in design and fabrication of micro-supercapacitors, Energy Environ Sci, 2014;7:867–884. doi:10.1039/C3EE43526A

[7] Z.L. Wang, Self-powered nanosensors and nanosystems, Adv Mater, 2012;24:280–285. doi:10.1002/adma.201102958

[8] J.-H. Sung, S.-J. Kim, S.-H. Jeong, E.-H. Kim, K.-H. Lee, Flexible micro-supercapacitors, J Power Sources, 2006;162:1467–1470. doi:10.1016/j.jpowsour.2006.07.043

[9] I. Hyun Jin, S. Kumar, S.-H. Yang, G. Barbastathis, Origami fabrication of nanostructured, three-dimensional devices: electrochemical capacitors with carbon electrodes, Appl Phys Lett, 2006;88:083104. doi:10.1063/1.2177639

[10] Y.Q. Jiang, Q. Zhou, L. Lin, Planar MEMS supercapacitor using carbon nanotube forests, in: 2009 IEEE 22nd International Conference on Micro Electro Mechanical Systems. MEMS 2009, 2009, pp. 587–590. doi:10.1109/memsys.2009.4805450

[11] D. Pech, M. Brunet, P.-L. Taberna, P. Simon, N. Fabre, F. Mesnilgrente, V. Conedera, H. Durou, Elaboration of a microstructured inkjet-printed carbon electrochemical capacitor, J Power Sources, 2010;195:1266–1269. doi:10.1016/j.jpowsour.2009.08.085

[12] J. Chmiola, C. Largeot, P.-L. Taberna, P. Simon, Y. Gogotsi, Monolithic carbide-derived carbon films for micro-supercapacitors, Science, 2010;328:480–483. doi:10.1126/science.1184126

[13] D. Pech, M. Brunet, H. Durou, P. Huang, V. Mochalin, Y. Gogotsi, P.-L. Taberna, P. Simon, Ultrahigh-power micrometre-sized supercapacitors based on onion-like carbon, Nat Nanotechnol, 2010;5:651–654. doi:10.1038/nnano.2010.162

[14] J. Feng, X. Sun, C. Wu, L. Peng, C. Lin, S. Hu, J. Yang, Y. Xie, Metallic few-layered VS2 ultrathin nanosheets: high two-dimensional conductivity for in-plane supercapacitors, J Am Chem Soc, 2011;133:17832–17838. doi:10.1021/ja207176c

[15] W. Gao, N. Singh, L. Song, Z. Liu, A.L.M. Reddy, L. Ci, R. Vajtai, Q. Zhang, B. Wei, P.M. Ajayan, Direct laser writing of micro-supercapacitors on hydrated graphite oxide films, Nat Nanotechnol, 2011;6:496–500. doi:10.1038/nnano.2011.110

[16] K. Wang, W. Zou, B. Quan, A. Yu, H. Wu, P. Jiang, Z. Wei, An all-solid-state flexible micro-supercapacitor on a chip, Adv Energy Mater, 2011;1:1068–1072. doi:10.1002/aenm.201100488

[17] M.F. El-Kady, R.B. Kaner, Scalable fabrication of high-power graphene micro-supercapacitors for flexible and on-chip energy storage, Nat Commun, 2013;4:1475–1475. doi:10.1038/ncomms2446

[18] S. Makino, Y. Yamauchi, W. Sugimoto, Synthesis of electro-deposited ordered mesoporous RuOx using lyotropic liquid crystal and application toward micro-supercapacitors, J Power Sources, 2013;227:153–160. doi:10.1016/j.jpowsour.2012.11.032

[19] M. Beidaghi, Y. Gogotsi, Capacitive energy storage in micro-scale devices: recent advances in design and fabrication of micro-supercapacitors, Energy Environ Sci, 2014;7:867–884. doi:10.1039/C3EE43526A

[20] E.N. Hoffman, G. Yushin, T. El-Raghy, Y. Gogotsi, M.W. Barsoum, Micro and mesoporosity of carbon derived from ternary and binary metal carbides, Microporous Mesoporous Mater, 2008;112:526–532. doi:10.1016/j.micromeso.2007.10.033

[21] M. Heon, S. Lofland, J. Applegate, R. Nolte, E. Cortes, J.D. Hettinger, P.-L. Taberna, P. Simon, P. Huang, M. Brunet, Y. Gogotsi, Continuous carbide-derived carbon films with high volumetric capacitance, Energy Environ Sci, 2011;4:135–138. doi:10.1039/c0ee00404a

[22] P. Huang, M. Heon, D. Pech, M. Brunet, P.-L. Taberna, Y. Gogotsi, S. Lofland, J.D. Hettinger, P. Simon, Micro-supercapacitors from carbide derived carbon (CDC) films on silicon chips, J Power Sources, 2013;225:240–244. doi:10.1016/j.jpowsour.2012.10.020

[23] C. Shen, X. Wang, W. Zhang, F. Kang, A high-performance three-dimensional micro supercapacitor based on self-supporting composite materials, J Power Sources, 2011;196:10465–10471. doi:10.1016/j.jpowsour.2011.08.007

[24] F. Liu, A. Gutes, C. Carraro, J. Chu, R. Maboudian, Graphitization of n-type polycrystalline silicon carbide and its application for micro-supercapacitors, in: 2011 16th International Solid-State Sensors, Actuators and Microsystems Conference, 2011, pp. 1879–1882.

[25] L. Wei, N. Nitta, G. Yushin, Lithographically patterned thin activated carbon films as a new technology platform for on-chip devices, ACS Nano, 2013;7:6498–6506. doi: 10.1021/nn4028129

[26] X. Yang, C. Cheng, Y. Wang, L. Qiu, D. Li, Liquid-mediated dense integration of graphene materials for compact capacitive energy storage, Science, 2013;341:534–537.

[27] T. Lin, I.W. Chen, F. Liu, C. Yang, H. Bi, F. Xu, F. Huang, Nitrogen-doped mesoporous carbon of extraordinary capacitance for electrochemical energy storage, Science, 2015;350:1508–1513. doi:10.1126/science.aab3798

[28] X. Cui, R. Lv, R.U.R. Sagar, C. Liu, Z. Zhang, Reduced graphene oxide/carbon nanotube hybrid film as high performance negative electrode for supercapacitor, Electrochimica Acta, 2015;169:342–350. doi:10.1016/j.electacta.2015.04.074

[29] D. Pech, M. Brunet, H. Durou, P. Huang, V. Mochalin, Y. Gogotsi, P.-L. Taberna, P. Simon, Ultrahigh-power micrometre-sized supercapacitors based on onion-like carbon, Nat Nano, 2010;5:651–654. http://www.nature.com/nnano/journal/v5/n9/abs/nnano.2010.162.html#supplementary-information

[30] J.J. Yoo, K. Balakrishnan, J. Huang, V. Meunier, B.G. Sumpter, A. Srivastava, M. Conway, A.L.M. Reddy, J. Yu, R. Vajtai, P.M. Ajayan, Ultrathin planar graphene supercapacitors, Nano Lett., 2011;11:1423–1427. 10.1021/nl200225J

[31] M. Kaempgen, C.K. Chan, J. Ma, Y. Cui, G. Gruner, Printable thin film supercapacitors using single-walled carbon nanotubes, Nano Lett, 2009;9:1872–1876. doi:10.1021/nl8038579

[32] C. Portet, G. Yushin, Y. Gogotsi, Electrochemical performance of carbon onions, nanodiamonds, carbon black and multiwalled nanotubes in electrical double layer capacitors, Carbon, 2007;45:2511–2518. doi:10.1016/j.carbon.2007.08.024

[33] M. Beidaghi, C. Wang, Micro-supercapacitors based on interdigital electrodes of reduced graphene oxide and carbon nanotube composites with ultrahigh power handling performance, Adv Funct Mater, 2012;22:4501–4510. doi:10.1002/adfm.201201292

[34] Z.S. Wu, K. Parvez, X. Feng, K. Müllen, Graphene-based in-plane micro-supercapacitors with high power and energy densities, Nat Commun, 2013;4. doi:10.1038/ncomms3487

[35] J.H. Lim, D.J. Choi, H.K. Kim, W.I. Cho, Y.S. Yoon, Thin film supercapacitors using a sputtered RuO_2 electrode, J Electrochem Soc, 2001;148:A275–A278. doi: 10.1149/1.1350666

[36] C.-C. Liu, D.-S. Tsai, D. Susanti, W.-C. Yeh, Y.-S. Huang, F.-J. Liu, Planar ultracapacitors of miniature interdigital electrode loaded with hydrous RuO_2 and RuO_2 nanorods, Electrochim Acta, 2010;55:5768–5774. doi:10.1016/j.electacta.2010.05.015

[37] J.P. Zheng, T.R. Jow, Q.X. Jia, X.D. Wu, Proton insertion into ruthenium oxide film prepared by pulsed laser deposition, J Electrochem Soc, 1996;143:1068–1070. doi: 10.1149/1.1836584

[38] B.E. Conway, Transition from "Supercapacitor" to "Battery" behavior in electrochemical energy storage, J Electrochem Soc, 1991;138:1539–1548. doi:10.1149/1.2085829

[39] T.M. Dinh, A. Achour, S. Vizireanu, G. Dinescu, L. Nistor, K. Armstrong, D. Guay, D. Pech, Hydrous RuO_2/carbon nanowalls hierarchical structures for all-solid-state ultrahigh-energy-density micro-supercapacitors, Nano Energy, 2014;10:288–294. doi: 10.1016/j.nanoen.2014.10.003

[40] C. Shen, X. Wang, W. Zhang, F. Kang, Direct prototyping of patterned nanoporous carbon: a route from materials to on-chip devices, SciRep, 2013;3:2294. doi:10.1038/srep02294

[41] J.H. Pikul, H. Gang Zhang, J. Cho, P.V. Braun, W.P. King, High-power lithium ion microbatteries from interdigitated three-dimensional bicontinuous nanoporous electrodes, Nat Commun, 2013;4:1732. doi:10.1038/ncomms2747

[42] H.-S. Min, B.Y. Park, L. Taherabadi, C. Wang, Y. Yeh, R. Zaouk, M.J. Madou, B. Dunn, Fabrication and properties of a carbon/polypyrrole three-dimensional microbattery, J Power Sources, 2008;178:795–800. doi: 10.1016/j.jpowsour.2007.10.003

[43] M. Kotobuki, Y. Suzuki, H. Munakata, K. Kanamura, Y. Sato, K. Yamamoto, T. Yoshida, Effect of sol composition on solid electrode/solid electrolyte interface for all-solid-state lithium ion battery, Electrochimica Acta, 2011;56:1023–1029. doi:10.1016/j.electacta.2010.11.008

[44] C. Shen, X. Wang, S. Li, J.g. Wang, W. Zhang, F. Kang, A high-energy-density micro supercapacitor of asymmetric MnO_2-carbon configuration by using micro-fabrication technologies, J Power Sources, 2013;234:302–309. doi:10.1016/j.jpowsour.2012.10.101

[45] W. Zhou, X. Cao, Z. Zeng, W. Shi, Y. Zhu, Q. Yan, H. Liu, J. Wang, H. Zhang, One-step synthesis of Ni_3S_2 nanorod@Ni(OH)(2) nanosheet core-shell nanostructures on a three-dimensional graphene network for high-performance supercapacitors, Energy Environ Sci, 2013;6:2216–2221. doi:10.1039/c3ee40155c

[46] H.Y. Lee, J.B. Goodenough, Supercapacitor behavior with KCl electrolyte, J Solid State Chem, 1999;144:220–223. doi:10.1006/jssc.1998.8128

[47] P. Simon, Y. Gogotsi, Materials for electrochemical capacitors, Nat Mater, 2008;7:845–854.

[48] S.C. Pang, M.A. Anderson, T.W. Chapman, Novel electrode materials for thin-film ultracapacitors: comparison of electrochemical properties of sol-gel-derived and electrodeposited manganese dioxide, J Electrochem Soc, 2000;147:444–450. doi: 10.1149/1.1393216

[49] C.-C. Lin, P.-Y. Lin, Capacitive manganese oxide thin films deposited by reactive direct current sputtering, J Electrochem Soc, 2010;157:A753–A759. doi:10.1149/1.3414823

[50] D.F. Yang, Pulsed laser deposition of manganese oxide thin films for supercapacitor applications, J Power Sources, 2011;196:8843–8849. doi:10.1016/j.jpowsour.2011.06.045

[51] I. Zhitomirsky, J. Wei, N. Nagarajan, Manganese oxide films for electrochemical supercapacitors, J Mater Process Technol, 2007;186:356–361. doi:10.1016/j.jmatprotec.2007.01.003

[52] W. Si, C. Yan, Y. Chen, S. Oswald, L. Han, O.G. Schmidt, On chip, all solid-state and flexible micro-supercapacitors with high performance based on MnOx/Au multilayers, Energy Environ Sci, 2013;6:3218–3223. doi:10.1039/C3EE41286E

[53] M. Nakayama, A. Tanaka, Y. Sato, T. Tonosaki, K. Ogura, Electrodeposition of manganese and molybdenum mixed oxide thin films and their charge storage properties, Langmuir, 2005;21:5907–5913. doi:10.1021/la050114u

[54] M. Nakayama, K. Suzuki, K. Okamura, R. Inoue, L. Athouel, O. Crosnier, T. Brousse, Doping of cobalt into multilayered manganese oxide for improved pseudocapacitive properties, J Electrochem Soc, 2010;157:A1067–A1072. doi:10.1149/1.3467865

[55] D.F. Yang, Pulsed laser deposition of vanadium-doped manganese oxide thin films for supercapacitor applications, J Power Sources, 2013;228:89–96. doi:10.1016/j.jpowsour.2012.11.067

[56] S.K. Meher, G.R. Rao, Effect of microwave on the nanowire morphology, optical, magnetic, and pseudocapacitance behavior of Co3O4, J Phys Chem C, 2011;115:25543–25556. doi:10.1021/jp209165v

[57] L. Gu, Y. Wang, R. Lu, W. Wang, X. Peng, J. Sha, Silicon carbide nanowires@Ni(OH)(2) core-shell structures on carbon fabric for supercapacitor electrodes with excellent rate capability, J Power Sources, 2015;273:479–485. doi:10.1016/j.jpowsour.2014.09.113

[58] G.Q. Zhang, H.B. Wu, H.E. Hoster, M.B. Chan-Park, X.W. Lou, Single-crystalline $NiCo_2O_4$ nanoneedle arrays grown on conductive substrates as binder-free electrodes for high-performance supercapacitors, Energy Environ Sci, 2012;5:9453–9456. doi:10.1039/C2EE22572G

[59] C. Liu, Z. Li, Z. Zhang, MoOx thin films deposited by magnetron sputtering as an anode for aqueous micro-supercapacitors, Sci Technol Adv Mater, 2013;14. 065005. doi: 10.1088/1468-6996/14/6/065005

[60] C. Liu, Z. Li, Z. Zhang, Molybdenum oxide film with stable pseudocapacitive property for aqueous micro-scale electrochemical capacitor, Electrochimica Acta, 2014;134:84–91. doi: 10.1016/j.electacta.2014.04.115

[61] E. Eustache, R. Frappier, R.L. Porto, S. Bouhtiyya, J.-F. Pierson, T. Brousse, Asymmetric electrochemical capacitor microdevice designed with vanadium nitride and nickel oxide thin film electrodes, Electrochem Commun, 2013;28:104–106.

[62] C. Liu, Z. Xie, W. Wang, Z. Li, Z. Zhang, Fabrication of MoOx film as a conductive anode material for micro-supercapacitors by electrodeposition and annealing, J Electrochem Soc, 2014;161:A1051–A1057. doi:10.1149/2.081406jes

[63] C. Liu, Z. Xie, W. Wang, Z. Li, Z. Zhang, The Ti@MoOx nanorod array as a three dimensional film electrode for micro-supercapacitors, Electrochem Commun, 2014;44:23–26. doi:10.1016/j.elecom.2014.04.007

[64] V. Presser, L. Zhang, J.J. Niu, J. McDonough, C. Perez, H. Fong, Y. Gogotsi, Flexible nano-felts of carbide-derived carbon with ultra-high power handling capability, Adv Energy Mater, 2011;1:423–430. doi:10.1002/aenm.201100047

[65] Z. Niu, L. Zhang, L. Liu, B. Zhu, H. Dong, X. Chen, All-solid-state flexible ultrathin micro-supercapacitors based on graphene, Adv Mater (Deerfield Beach, Fla.), 2013;25:4035–4042. doi:10.1002/adma.201301332

[66] B. Hsia, M.S. Kim, M. Vincent, C. Carraro, R. Maboudian, Photoresist-derived porous carbon for on-chip micro-supercapacitors, Carbon, 2013;57:395–400. doi:10.1016/j.carbon.2013.01.089

[67] S.W. Lee, B.-S. Kim, S. Chen, Y. Shao-Horn, P.T. Hammond, Layer-by-layer assembly of all carbon nanotube ultrathin films for electrochemical applications, J Am Chem Soc, 2009;131:671–679. doi:10.1021/ja807059k

[68] C. Arbizzani, M. Mastragostino, L. Meneghello, Polymer-based redox supercapacitors: a comparative study, Electrochimica Acta, 1996;41:21–26. doi: 10.1016/0013-4686(95)00289-Q

[69] J.C. Carlberg, O. Inganäs, Poly(3,4-ethylenedioxythiophene) as electrode material in electrochemical capacitors, J Electrochem Soc, 1997;144:L61–L64. doi:10.1149/1.1837553

[70] W. Sun, R. Zheng, X. Chen, Symmetric redox supercapacitor based on micro-fabrication with three-dimensional polypyrrole electrodes, J Power Sources, 2010;195:7120–7125. doi:10.1016/j.jpowsour.2010.05.012

[71] M. Beidaghi, C. Wang, Micro-supercapacitors based on three dimensional interdigital polypyrrole/C-MEMS electrodes, Electrochimica Acta, 2011;56:9508–9514. doi:10.1016/j.electacta.2011.08.054

[72] D.R. Rolison, J.W. Long, J.C. Lytle, A.E. Fischer, C.P. Rhodes, T.M. McEvoy, M.E. Bourg, A.M. Lubers, Multifunctional 3D nanoarchitectures for energy storage and conversion, Chem Soc Rev, 2009;38:226–252. doi:10.1039/b801151f

Chapter 3

Engineering Nanostructured MnO_2 for High Performance Supercapacitors

Jian-Gan Wang

Additional information is available at the end of the chapter

http://dx.doi.org/10.5772/65008

Abstract

Manganese oxides (MnO_2) have particularly received increasing attention owing to their high theoretical specific capacitance of 1370 F/g, low-cost, natural abundance, and environmental benignity. However, MnO_2 suffers from low electrical conductivity (10^{-5} to 10^{-6} S/cm), low ionic diffusion constant (~10^{-13} cm^2/V s), and low structural stability, which results in low electrochemical utilization and poor cycling life. It is therefore important to explore new strategies to improve the electrochemical performance of MnO_2. The effective methods to maximize the performance involve (i) reducing MnO_2 structures to a nanoscale range and (ii) compositing MnO_2 with highly conductive materials. In this chapter, we will first introduce the rapid development of MnO_2 nanostructures for supercapacitors. Then the fundamental charge storage mechanism of MnO_2 will be specifically clarified. The preparation methods of MnO_2 nanostructures and their composites will be subsequently summarized. Then, we will pay great attention to the most recent development of MnO_2-based nanostructures for supercapacitors, which is the main body of this chapter. The practical application of MnO_2 nanostructures for symmetric and asymmetric supercapacitors will be discussed. Finally, we will present a brief perspective regarding the rational design and synthesis of MnO_2-based nanostructures.

Keywords: MnO_2, nanostructure, composite, supercapacitor, high performance

1. Introduction

According to the charge storage mechanism, supercapacitors can be categorized as electrochemical double layer capacitors (EDLCs) and pseudocapacitors [1]. The typical electroactive electrode materials for pseudocapacitors include transition-metal oxides and conducting polymers. To this end, massive efforts have been dedicated to developing low-cost and eco-friendly alternatives, such as MnO_2, Co_3O_4, and NiO. Among the alternative candidates, MnO_2 has received intensive attention in the past decades. Lee and Goodenough reported the first use of MnO_2 for supercapacitors in 1999 [2]. This finding paved a new avenue for exploring a huge variety of electroactive MnO_2-based electrode materials for high performance supercapacitors. This is because MnO_2 shows advantages of (i) a high theoretical specific capacitance of 1370 F/g; (ii) a wide operating potential window of about 1.0 V; and (iii) the ability to enable mild aqueous electrolytes with a much less chemical corrosion to current collectors or packages [3, 4]. In addition, MnO_2 has additional merits of natural abundance, low-cost, and environmentally benignity [4]. These unique characteristics will enable MnO_2-based electrodes to act as a high-performance, safe, and a low-cost replacement.

However, the specific capacitance of MnO_2 is almost one magnitude lower than the theoretical value [4–6]. This is because MnO_2 shows poor electronic conductivity (10^{-5} to 10^{-6} S/cm), low ionic diffusion constant (~10^{-13} cm^2/V s), and structure flexibility [7, 8]. Moreover, owing to a wide diversity of textural feature, crystal forms (e.g., α-, β-, δ-, γ-MnO_2), and defect chemistry, MnO_2 exhibits a variety of distinct electrochemical performance. The performance of MnO_2 will be substantially improved with a better fundamental understanding of the charge storage mechanism at the electrode/electrolyte interfaces and a rational design of MnO_2-based electrodes.

In recent years, extensive endeavors have been dedicated to synthesizing MnO_2 nanostructures or nanocomposites with different morphologies and crystal forms. This chapter will introduce the state-of-the-art MnO_2 electrodes for high performance supercapacitors. The content includes a brief discussion of the charge storage mechanisms of MnO_2, a summary of the synthetic methodologies, a main focus on the MnO_2-based electrodes and their application in supercapacitors.

2. Charge storage mechanism of MnO_2

The charge storage mechanism of MnO_2 was first imitated from RuO_2 and Zn/MnO_2 cells in acidic electrolytes [9], which involve a reversible insertion/extraction of protons into/from MnO_2 with a concomitant Mn^{3+}–Mn^{4+} transition. However, the specific capacitance of MnO_2 is found to be related to the species and concentrations of the alkali metal cations, such as Li^+, Na^+, and K^+, regardless of the pH value [10, 11]. Consequently, a more rational mechanism based on the chemisorption of alkali metal cations on the surface of MnO_2 was proposed and can be written as [3, 11]:

$$\left(MnO_2\right)_{surface} + M^+ + e^- \rightarrow \left(MnOOM\right)_{surface} \tag{1}$$

where M^+ represents alkali metal cations, including Li^+, Na^+, and K^+. Later, Belanger and co-workers developed a new complementary storage mechanism, that is, the alkali metal cations can insert/extract into/from MnO_2, as shown in Eq. (2) [11]:

$$MnO_2 + M^+ + e^- \rightarrow MnOOM \qquad (2)$$

It should be noted that MnO_2 can crystallize in various forms. Can the above charge storage mechanisms be suitable for all MnO_2 structures? Brousse et al. investigated a systematic comparison on the supercapacitive properties of MnO_2 with five types of phase structures [11]. It is revealed that the capacitance is closely associated with the tunnel size of crystalline MnO_2, i.e., the larger the tunnel size, the larger the specific capacitance. More specifically, the one-dimensional (1D) α-MnO_2 (4.6 Å) and two-dimensional (2D) δ-MnO_2 (7 Å) allow a fast insertion of hydrated K^+ cations (3 Å) due to their large tunnel sizes, while β-MnO_2 and γ-MnO_2 with one-dimensional tunnel sizes smaller than cations limit the diffusion process, and the spinel λ-MnO_2 with a more opened three-dimensional (3D) structure can permit a partial cationic diffusion. This suggests that the tunnel space should be large enough to allow a high-rate insertion/extraction of electrolyte ions for charge storage governed by Eq. (2), which is known as "tunnel storage mechanism (TSM)" [12]. Hence, the specific capacitance of MnO_2 depends strongly on its crystallographic structure, which follows the decreasing order: $\alpha(m) > \alpha \approx \delta > \gamma > \lambda > \beta$. Furthermore, owing to the large tunnel, α-MnO_2 also exhibits ideal pseudocapacitive behavior in the electrolytes containing Ca^{2+}, Mg^{2+}, or Ba^{2+}, which can store more charge through a multivalent cation storage mechanism [5]. In addition to the metal cations, anions can also be the working species that compensate the Mn-valent state variation upon charge/discharge in aprotic ionic liquids (IL) [13]. In summary, the charge/discharge process of MnO_2 typically involves (i) electrolyte ion transport; (ii) ion adsorption/desorption at the surface sites of electrodes, which are dependent on the ion size, the mobility of ion, and the dehydration/hydration rate; and (iii) ion insertion/extraction into/from bulk MnO_2. Therefore, the charge storage process is not only associated with crystallographic structures, but also relates to other factors, such as specific surface area, electronic conductivity, and ionic conductivity, etc. Therefore, the charge storage mechanism of MnO_2 basically involves a capacitive surface chemisorption process (Eq. (1)) and a bulk/subsurface Faradaic reaction (Eq. (2)).

Recently, mixed-valent chemistry of metal oxides charge compensation mechanisms at oxygen centers and surfaces were proposed as a new charge storage mechanism. The mixed-valent MnO_x nanostructures exhibited a high-specific capacitance of 2530 F/g in a two-electrode configuration, about double of the theoretical value (1370 F/g) [14]. It is revealed that a large portion of charge compensation might be originated from the hole state redistribution toward the O sites, rather than merely reduction of Mn ions. In addition, the ionic defects (vacancies and misplaced ions), electronic defects (electron and holes), and structural defects (cavities, stacking faults, etc.) in the mixed-valent MnO_x can also boost the kinetics of the surface redox reactions and the transport of charged species. These findings may offer critical insights into the fundamental understanding of the relationship between mixed-valent structures and electrochemical properties and to rational design of a new-generation supercapacitors.

3. Synthesis methods

3.1. Chemical precipitation

MnO_2 nanostructures can be easily prepared *via* various chemical co-precipitation methods. The most commonly one is based on a simple redox reaction between MnO_4^- and Mn^{2+} [4]:

$$2MnO_4^- + 3Mn^{2+} + 2H_2O \rightarrow 5MnO_2 + 4H^+ \tag{3}$$

Amorphous hydrated MnO_2 powders are the resulting products at room temperature. Increasing the reaction temperature not only leads to an increase in the degree of MnO_2 crystallinity, but also generates different morphologies, such as nanorods, nanoflowers, and nanotubes [15]. Because of the strong oxidation of MnO_4^-, a large number of reducing species, including acid, polymer monomers, carbon, etc., can be used to obtaining MnO_2 nanostructures [16]. In addition, MnO_2 can also be gained by changing other oxidants, such as $Na_2S_2O_8$, $KClO_3$, etc. These flexible derived reactions facilitate the formation of MnO_2-carbon and MnO_2-conducting polymer nanocomposites [17].

3.2. Electrodeposition

3.2.1. Anodic electrodeposition

Anodic electrodeposition of MnO_2 nanostructures involves an oriented diffusion of Mn^{2+} species to the anode surface under an applied electric field and a subsequent electro-oxidation of the species into MnO_2. The anodic deposition reaction can be expressed as [18]:

$$Mn^{2+} + 2H_2O \rightarrow MnO_2 + 4H^+ + 2e^- \tag{4}$$

Three different deposition techniques have been developed: i.e., potentiostatic deposition, potentiodynamic deposition, and galvanostatic deposition. The morphology can be well controlled by modifying the deposition parameters (such as applied potentials, applied currents, scan rates, electrolyte ingredients/concentrations, and pH value, etc.) and by tuning the physiochemical nature of substrates (such as a porous structure, a hydrophilic surface, etc.).

In addition to pure MnO_2, conducting polymer-MnO_2 nanocomposites can also be fabricated using solution precursors containing both Mn^{2+} species and monomers [19]. One-step or two-step electrochemical co-deposition has been explored to effectively prepare a nanocomposite. The co-electrodeposition process primarily involves the anodic deposition of MnO_2 and the simultaneous electropolymerization of conducting polymers. The morphology of the nano-structured hybrids can be controlled by modifying the deposition parameters.

3.2.2. Cathodic electrodeposition

Cathodic electrodeposition of MnO_2 nanostructures can take place at the cathode surface *via* two pathways based on Mn species in solution under a negative potential. The first way is based on the electrochemical processes in aqueous solution containing Mn^{2+} species, which include water electrolysis and oxygen reduction to generate OH^-, followed by the precipitation of metastable $Mn(OH)_2$ and a further oxidation into MnO_2. The other pathway is on basis of the reduction of Mn^{7+} species from MnO_4^- on the cathode surface, following reaction (5) [20]. The MnO_2 nanostructures can be shape-controlled by tuning deposition variables of current density, potential, solution concentration, and pH value.

$$MnO_4^- + 2H_2O + 3e^- \rightarrow MnO_2 + 4OH^- \tag{5}$$

3.3. Electrostatic interaction assembly

Electrostatic interaction assembly is an effective technique for fabricating nanocomposites *via* a self-assembly of oppositely charged materials. The strong electrostatic attractive interactions between positively charged species and negatively charged species can ensure the robustness of resulting materials. The binding intensity of the electric force is determined by the Coulomb's law [21]:

$$F = q_+ q_- / r^2 \tag{6}$$

where F is the electric force, q_+ and q_- are the charges of the negative and positive species, and r is the distance between the two species. The main challenge is to controllably modify the target material surfaces with negatively/positively charged species, such as negatively charged poly(styrene sulfonate) (PSS), and positively charged aminopropyltrimethoxysilane (APTS) moieties and poly(diallyldimethylammonium chloride) (PDDA) [22].

4. Electrochemical performance of pure MnO_2 nanostructures

4.1. Amorphous MnO_2 powder electrodes

The MnO_2 powders prepared at low temperatures are generally amorphous or poorly crystalline in nature [23]. In addition, the as-prepared samples often contain a certain amount of hydrated content, which exists in the form of residual structural water and/or hydrated trivalent MnOOH. The hydrated MnO_2 powders have large surface area because of ultrafine nanoparticle morphology. Although the amorphous structure can be maintained up to 300°C, the hydrated amount and the surface area decrease significantly with increasing annealing temperature and dwelling time [8, 24].

The electrochemical properties of the amorphous MnO_2 powder electrode are closely related to their surface area and hydrated amount. According to the Eq. (1), an increase in the surface area is rather favorable for enhancing specific capacitance. The hydrated species can also modify the MnO_2 surface affinity, which facilitates the electrolyte access and ion transport for enhanced performance. It was revealed that the amorphous MnO_2 had the largest surface area and also possessed the best electrochemical performance. A dramatic decrease in the surface area and water content were observed after annealing treatments, which, in turn, degraded the specific capacitance.

4.2. Crystalline MnO_2 powder electrodes

As mentioned earlier, MnO_2 can exist in various crystal forms. The arrangement of [MnO_6] basic building unit enables the construction of one-dimensional, two-dimensional, and three-dimensional tunnel structures. The tunnel space, depending on the tunnel sizes, can accommodate cations, such as K^+, Na^+, Li^+, Ca^+, Mg^+, etc., which favors the energy storage according to Eq. (2). Recently, Ghodbane et al. has carried out a systematic study dealing with the microstructural effects on electrochemical properties in crystalline MnO_2 [7]. As shown in **Figure 1**, the three-dimensional spinel λ-MnO_2 exhibits the highest specific capacitance followed by the two-dimensional δ-MnO_2 and, finally, the one-dimensional group shows the lowest values. To some extent, the specific capacitance increases with an increase in the tunnel size. One exception is the Ni-todorokite, which has a large tunnel size but a low-specific capacitance. The reason is the presence of very short and strong hydrogen bonds, which stabilize the hydrated cations inside the cavity and impede the ion diffusion through the tunnels. Additionally, it was found the specific capacitance did not depend on the specific surface area, and the ionic conductivity exerted great influence on the electrochemical performance. The comparison suggests the charge storage in MnO_2 is a fundamentally Faradaic process.

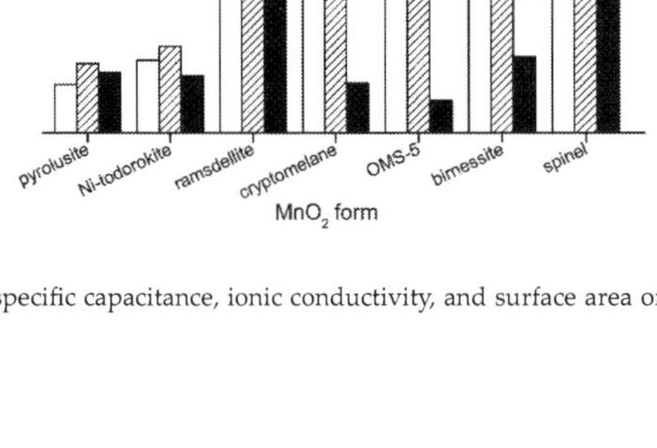

Figure 1. Comparison of the specific capacitance, ionic conductivity, and surface area of various MnO_2 nanostructures [7].

Hydrothermal or solvothermal methods are appropriate techniques to prepare MnO_2 nanostructures, including nanorods, nanotubes, and nanowires, [4, 15]. The as-prepared products showed distinct morphologies, but the electrochemical results are highly scattered. There is no consistent relationship among the synthesis conditions, microstructures, and specific capacitance. In general, MnO_2 with high crystallinity shows a low-specific surface area, and the cations in the tunnel structure are difficult to be extracted for TSM, thereby resulting in limited specific capacitance.

4.3. Thin-film MnO_2 electrodes

Thin-film MnO_2 electrodes have been massively explored for deepening fundamental studies and finding potential applications as microscale energy storage devices. In this regard, a number of MnO_2 thin-film electrodes have been directly prepared on a current collector, such as metal foils, carbon cloth, etc., through anodic/cathodic electrodeposition. In terms of the charge storage mechanism, the most desirable morphology should be of three-dimensional and porous nanoarchitectures with good connection to a highly conductive substrate. Therefore, much effort has focused on the morphology-controlled growth of porous MnO_2 nanostructures, with a purpose of obtaining more accessible electroactive sites and short ion diffusion pathways. This can be achieved by controlling the deposition conditions and using porous templates. For example, galvanostatic, or potentiostatic techniques were employed to fabricate porous fibrous network, nanorods, and nanosheets [25–27].

As for anodic electrodeposition, chemically stable metal (e.g., Ti or Ta foils, etc.) and carbon (graphites, carbon fabric, etc.) substrates should be employed to address the problem of anodic oxidation and dissolution of metal substrates (e.g., stainless steel). In sharp contrast, cathodic deposition is a good method to break this limit. Electrodeposition from a dilute $NaMnO_4^-$ solution (0.02 M) produced a typical surface morphology of fibrous film with a birnessite-type crystal structure [28]. Compared to the pure MnO_2 powders in Sections 4.1 and 4.2, the MnO_2 thin films can deliver much higher specific capacitance.

4.4. Summary

Pure MnO_2 nanostructures with different crystal phases and morphologies can be easily prepared using solution-based methods. Unfortunately, the overall specific capacitances of the amorphous and crystalline MnO_2 powders are mostly ranging from 100 to 250 F/g, which are far from satisfaction. According to the charge storage mechanisms, the Faradaic reactions of MnO_2 only occur at the surface or subsurface within tens of nanometers. Thus, to maximize the high electrochemical utilization of MnO_2, the thickness of MnO_2 should be kept in a nanoscale range. This is confirmed by a high-specific capacitance of 300–800 F/g when MnO_2 thin films were deposited on the conductive substrates. The specific capacitance can even reach the theoretical value when thickness of MnO_2 film is only several nanometers. The reason can be rationalized by the nanostructured MnO_2 shortening both ionic and electronic transport/diffusion distances for faster reaction kinetics. However, the thin films are limited by their low mass loading, which cannot meet large scale practical applications.

5. Electrochemical performance of MnO_2-based nanocomposites

5.1. MnO_2-carbon nanocomposites

Over the past decades, carbon nanostructures have received intensive attention in various communities of material science and chemistry. First, carbon nanostructures are the most effective electrode material candidates for EDLCs owing to the versatile size dimensionality, high surface area, good electrical conductivity, and strong structural stability. Nevertheless, carbon-based electrodes are limited by their low specific capacitance. To break this limitation, carbon structures often serve as favorable supports for MnO_2 by making complementary functions. In this section, we emphasize on the composites with one-dimensional carbon nanotubes (CNTs) two-dimensional, graphene, and three-dimensional mesoporous carbon [29].

5.1.1. MnO_2-CNTs

CNTs exhibit fascinating physicochemical properties of high electrical conductivity, high surface area, good chemical stability, and strong mechanical strength [29]. CNTs are increasingly employed as substrates for anchoring pseudocapacitive MnO_2, to form nanocomposites. The nanocomposites can utilize the large pseudocapacitance of MnO_2 and the conducting and mechanical merits from CNTs. Moreover, the nanocomposites could inherit the well-developed CNTs networks with open mesopores. Consequently, the MnO_2-CNTs nanocomposites are expected to exhibit superior electrochemical properties. The key issue to obtain high-performance MnO_2/CNT nanocomposites is how to uniformly deposit nanostructured MnO_2 onto CNTs. To ensure the uniformity, the CNT powders should be homogeneously dispersed in a solution by means of surface modification and/or ultrasonic treatments. The three-dimensional CNT network in the suspension can combine MnO_2 nanostructures (e.g., nanorods or nanoclusters) using various chemical or electrochemical deposition methods. The most favorable method is based on a spontaneously redox reaction of MnO_4^- and carbon, enabling *in situ* deposition of MnO_2 on CNTs [30–32]. **Figure 2a** shows a typical core-sheath hierarchy architecture constructing by MnO_2 nanostructures surrounding the CNT surfaces [33]. Under assistances of some modified techniques, such as the hydrothermal method, the refluxing, and the microwave irradiation, a variety of MnO_2/CNT nanocomposites with different mass loadings and morphologies (e.g., nanoflakes and nanorods) of MnO_2 have been demonstrated to exhibit improved electrochemical performance.

In addition to the surface deposition, MnO_2 nanoparticles can also be embedded into the well-defined CNT nanochannels, as shown in **Figure 2b** [34]. The confined MnO_2 are likely to exist in a more reduced state, and accordingly, the pseudocapacitive performance of MnO_2 can be improved by modifying the redox couples of Mn^{3+}/Mn^{4+}. The specific capacitance normalized by MnO_2 is almost close to the theoretical value (1250 F/g). These intriguing results may provide new insights into the configuration of MnO_2 with nanocarbon for developing high-performance electrode materials.

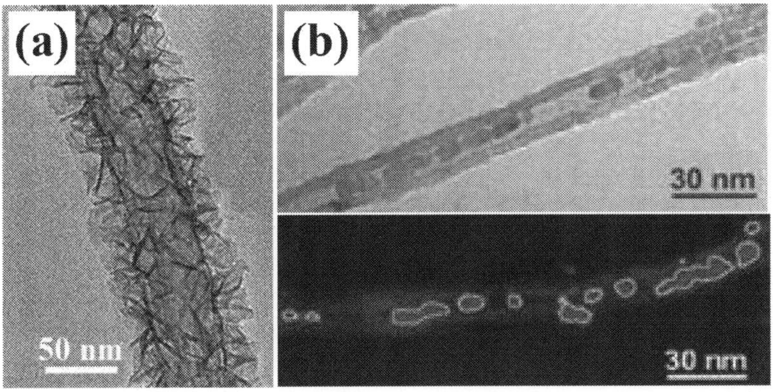

Figure 2. TEM (Transmission Electron Microscopy) images of (a) MnO_2-out-CNT and (b) MnO_2-in-CNT nanocomposites.

5.1.2. MnO_2-graphene

Graphene is a hottest star material in recent years due to its attractive characteristics of high electrical conductivity, good mechanical flexibility, high theoretical surface area (2600 m^2/g), and high thermal and chemical stability [35]. To exploit the potential of graphene-MnO_2 nanomaterials for supercapacitors, the main challenge is to prevent graphene nanosheets from restacking during material synthesis and cycling operation. MnO_2 nanostructures can act as interlayer spacers to effectively suppress the graphene restacking, thus facilitating fast ion diffusion/transport within the electrode materials. Moreover, the electrical intimate interaction of MnO_2 and graphene can boost interfacial charge transfer to ensure rapid redox reactions of MnO_2. In addition, graphene is capable of playing a "flexible confinement" role to enwrap MnO_2 nanostructures through preventing inner MnO_2 from electrochemical dissolution. Therefore, a strong synergetic effects benefiting from the enhanced conductivity, increased interfacial area as well as reinforced structural stability can be yielded in MnO_2-graphene nanocomposites. To date, a great number of studies are devoting to (i) addressing the synthesis complexity and scalability, (ii) tailoring MnO_2 nanostructures with a desired morphology and mass loadings in between graphene nanosheets, and (iii) improving electrical and mechanical connections between graphene and MnO_2.

There are two favorable approaches for the controlled fabrication of the target MnO_2/graphene nanocomposites. The first one is based on an *in situ* redox reaction between $KMnO_4$ and graphene or GO, as illustrated in **Figure 3a** [36]. The rich oxygen-containing functionalities of graphene layers ensure a high-MnO_2 mass loading. Functionalization of graphene, such as nitrogen doping, has become a key-enabling technology to improve the activities of graphene by increasing its conductivity and surface active sites. The other approach is a solution-based chemical mixing of separate MnO_2 nanostructures with graphene, i.e., the formation of MnO_2 is independent on graphene. Based on the electrostatic interaction assembly method, MnO_2

nanostructures can be uniformly incorporated and strong anchored in between the graphene nanosheets [22, 37]. **Figure 3b** and **c** exemplify a graphene-wrapped MnO_2 nanocomposite by co-assembling positively charged honeycomb-like MnO_2 nanospheres and negatively charged graphene nanosheets [22]. It is noted that the morphology, crystallinity, and mass loading of the nano-MnO_2 in the nanocomposites have great influence on the performance. In addition, controlling oxygen content of graphene plays a critical role in the fabrication of high-performance nanocomposites. It is believed that a simple, low-cost, and eco-friendly method is more favorable for controllably and scale-up fabricating MnO_2/graphene nanocomposites with high performance.

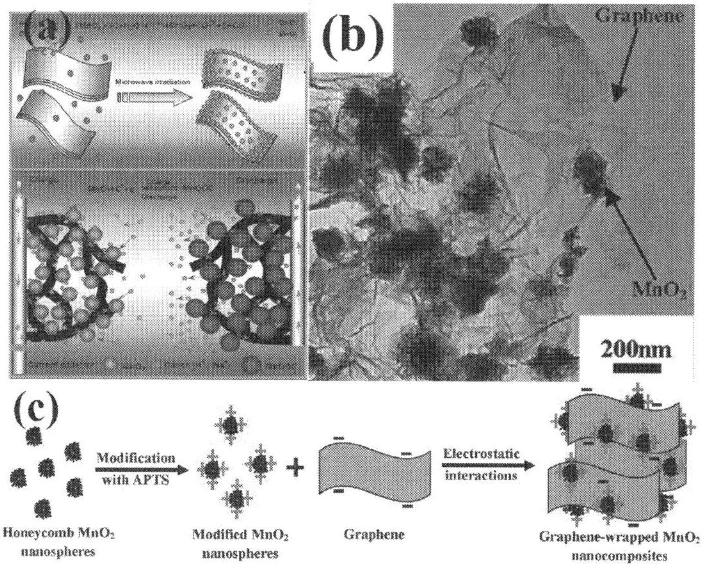

Figure 3. (a) Schematic illustration for redox deposition and charge storage process of MnO_2/graphene nanocomposite. (b) TEM image of MnO_2 honeycombs/graphene. (c) Schematic illustration for honeycomb-like MnO_2 on graphene through electrostatic assembly.

5.1.3. MnO_2/carbon nanotube/graphene

CNTs could serve as a spacer to prevent adjacent graphene nanosheets from restacking and as conductive networks to accelerate electron transport of hybrid carbon composites. Recently, a few studies have exploited MnO_2/CNT/graphene nanocomposites for high performance supercapacitors. The highly porous and conductive CNT/graphene composites can offer fast electronic and ionic channels for reversible Faradic redox reaction of MnO_2 nanostructures. MnO_2-coated CNTs were intercalated in between graphene nanosheets to form hierarchical nanocomposites [38]. The core step is the creation of positively charged MnO_2/CNT functionalized with PDDA, which drives an electrostatical self-assembly with highly negatively charged graphene nanosheets. In this architecture, MnO_2/CNT is able to effectively prevent graphene nanosheets from severe agglomeration. In addition to solution-based chemical assembly, chemical vapor deposition (CVD) method was recently utilized to grow CNT spacers in between graphene nanosheets [39]. The highly conductive network and the porous architecture enable the nanocomposites to exhibit an improvement in the specific capacitance.

5.1.4. MnO_2-mesoporous carbon

Compared to activated carbon, ordered mesoporous carbons (OMC) possess uniform mesopores of several nanometers in diameter. The presence of the ordered mesopores is beneficial for electrolyte wetting and rapid ionic motion, which could address the rate-limiting issue of supercapacitors. The favorable mesostructure also enables OMC to be an ideal host material for compositing MnO_2 nanostructures. However, the encapsulated MnO_2 may result in disappearance of uniform mesopores and generate micropores between the nanoparticles, which prevent the mass transfer of electrolyte ions and also the formation of a double-layer required for high performance supercapacitors [40]. Hence, controlled growth of the nanoparticles homogeneously within the mesopores is crucial.

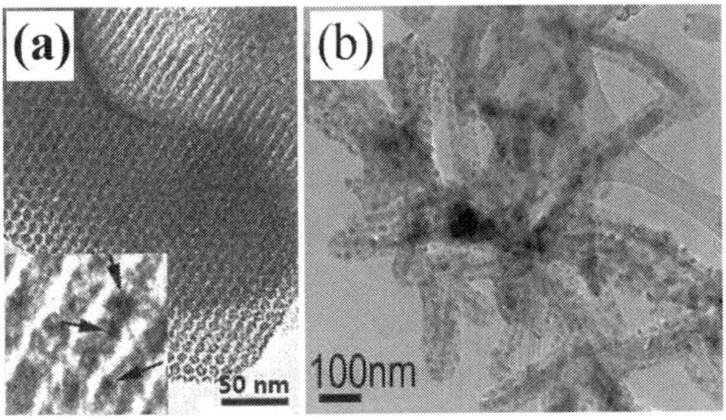

Figure 4. (a) TEM images of MnO_2/CMK-3. (b) TEM image of highly graphitic carbon-tipped MnO_x/mesoporous carbon/MnO_x hybrid nanowires.

Figure 4a exhibits a MnO_2/OMC composite prepared using an *in situ* redox reaction between $KMnO_4$ and the mesoporous carbon [41]. MnO_2 nanostructures are uniformly incorporated into the mesoporous carbon wall and the ordered mesopore structures are well-preserved. It is anticipated that OMC hosts are capable of holding the advantage of providing fast ion transport pathways for a high-rate power delivery; however, the rate capability of the composites is scarcely evaluated. This may be attributed to the relatively low electrical conductivity of OMC because the carbonization temperature is generally no more than 1000°C. This factor would become a bottle neck that would kinetically limit the charge transfer process. In combination of the favorable mesoporous structure and enhanced electrical conductivity, a novel one-dimensional, highly graphitic carbon-tipped MnO_x/mesoporous carbon/MnO_x (MMCM) hybrid nanostructure was demonstrated as a high-performance electrode material [42]. The unique TEM image of as-prepared hybrid nanowires are illustrated in **Figure 4b**, which is different from traditional structures with MnO_2 deposited on the surfaces of carbon materials. Such a highly graphitic carbon-tipped mesoporous carbon shell provides efficient channels for ion transport into the core MnO_x and improves electrical conductivity for electron transfer. The fascinating results offer a new direction on the design of ideal electrode materials with a high-specific capacitance, an excellent rate capability, and a long-term cyclability.

5.2. MnO_2-conducting polymer nanocomposites

Conducting polymers are another class of pseudocapacitive materials due to their fast and reversible doping/undoping kinetics, ease in preparation, environmental stability, and anti-corrosion purposes [43]. The conducting polymers could provide good electronic conductivity for MnO_2, and in turn, MnO_2 offers a solid support and a percolated electrical conducting pathway by interlinking the polymer chains, thus improving charge exchange efficiency and stability during redox cycling. In combination of high pseudocapacitance of both components, a strong synergistic effect is expected to integrate into a composite of MnO_2/conducting polymers.

5.2.1. MnO_2-polyainiline (PANi)

Polyaniline (PANi) is one of the most used conducting polymers due to its high doping level, good electrical conductivity, and environmental stability. Theoretical specific capacitance of PANi reaches 750 F/g, which makes it being widely used to combine various MnO_2 [56]. $KMnO_4$ is the common oxidant to polymerize aniline monomers into PANi accompanying with an instantaneous formation of MnO_2:

$$KMnO_4 + \text{aniline} \rightarrow MnO_2 + \text{PANi} \tag{7}$$

This method leads to a good contact at an inter-molecule level between each component. It is demonstrated mesoporous MnO_2/PANi hollow spheres using an interfacial synthesis technique (**Figure 5a**) [16]. The coupling reaction was carried out at the organic/water interfacial region, favoring the self-assembly of the composites with high surface area, uniform pore-size distribution, and hierarchical architecture. In addition to the aniline monomers, PANi nanostructures were used as reactive templates to reduce $KMnO_4$, with MnO_2 nanostructures depositing onto the PANi surface. **Figure 5b** shows ultrathin MnO_2 nanorods grown on surfaces of conducting polymer nanofibers (PANi, polypyrrole (PPy), and PEDOT (poly(3,4-ethylenedioxythiophene)) by simply soaking the nanofibers in a $KMnO_4$ aqueous solution [43]. Furthermore, owing to the high chemical oxidation potential of MnO_2 (1.23 V) in an acidic condition, the aniline monomers can be polymerized with a simultaneous reduction of MnO_2 into soluble Mn^{2+} ions, which provides a rational chemical method for *in situ* dispersing MnO_2 on PANi supports [44]. Owing to the reactive template nature, morphologies of the nanocomposites can be shaped by the pristine MnO_2, and the PANi content can be controlled by the polymerization time or the reaction ratio of aniline/MnO_2.

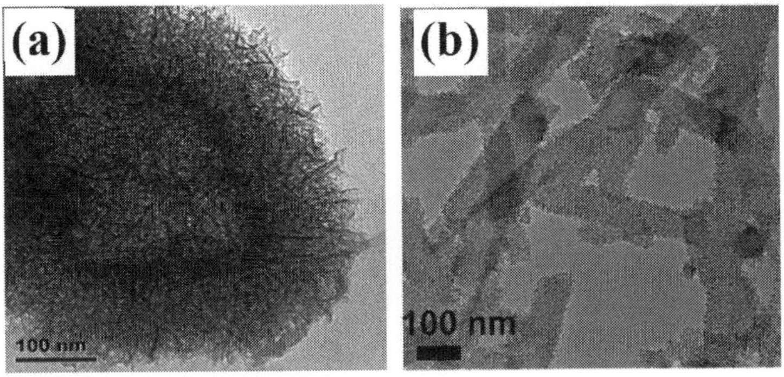

Figure 5. TEM images ofMnO_2/PANi (a) hollow spheres and (b) nanofibers.

5.2.2. MnO_2-polypyrrole (PPy)

PPy is another low-cost conducting polymer that can be used as promising electrode materials for supercapacitors. It is noteworthy that, unlike PANi showing good supercapacitive behavior in an acidic electrolyte, PPy holds the capability to possess excellent electrochemical properties in neutral electrolytes, which are highly compatible with MnO_2. $KMnO_4$ is a proper oxidant to chemically polymerize pyrrole monomers into PPy in relation to the formation of MnO_2 by reducing MnO_4^- [45]. The dispersed MnO_2 nanoparticles adhered to PPy chains increase the surface area and retard the structural deterioration of PPy backbones. As mentioned earlier, MnO_2 nanostructures could serve as solid reactive templates for *in situ* polymerization of pyrrole monomers in an acidic environment, thus forming MnO_2/PPy nanocomposites without using any surfactants and/or additional oxidants. To this end, unique one-dimensional coaxial MnO_2/PPy nanotubes were developed [46, 47].

5.2.3. MnO_2-polythiophone (PTh)

Lu and Zhou reported a one-pot interfacial synthesis to fabricate mesoporous MnO_2/PTh nanocomposite having uniform submicron-sphere/nanosheet hierarchical structures [48]. The significant roles of PTh in the nanocomposite are emphasized in terms of their functions on enhancing the electrical conductivity and constraining the dissolution of MnO_2 component.

5.3. MnO_2-carbon-conducting polymer ternary nanocomposites

In order to fully exploit the advantages of different carbon and/or conducting polymers, MnO_2 can be combined with the dual-supports to make a maximum use of its utilization. Carbon nanostructures (e.g., CNTs) can facilitate easy electrolyte accessibility and fast electron transport into the bulk electrode materials, meanwhile the conducting polymers can contribute to more charge storage and better inter-particle connectivity to MnO_2. This ternary design approach is expected to show improved performance, since each component in the ternary nanocomposite serves its unique and desired functions to collectively optimize their electrochemical properties.

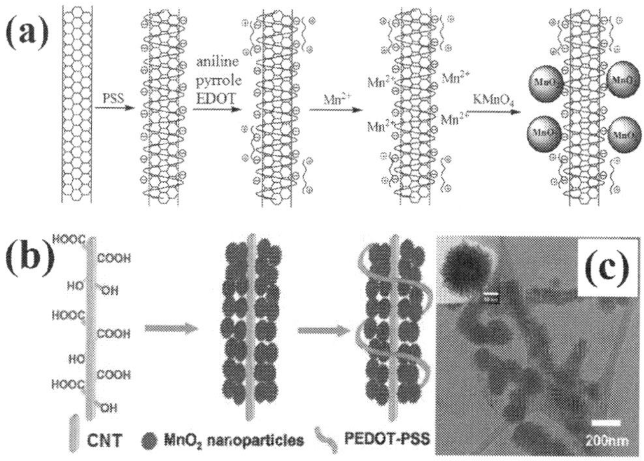

Figure 6. (a, b) Schematic illustration of synthesizing MnO_2/conducting polymers on CNTs-PSS. (c) TEM image of MnO_2/PEDOT/CNTs.

As for the MnO_2-CNT-conducting polymer nanocomposites, **Figure 6a** presents a typical synthesis route [21]. The CNT surface was modified by wrapping negatively charged PSS, which electrostatically attract Mn^{2+} species and monomers (i.e., aniline, pyrrole, and EDOT). The large number of surface sites ($-SO_3^-$) have strong interactions of growing nucleus, which facilitate an ordered growth of nanostructured MnO_2 and conducting polymers and further hinder the inter-particle agglomeration. The molecular level dispersion of MnO_2 in the CNT networks and conducting polymer matrixes results in a strong synergistic interaction. Electrochemical polymerization can also be used to deposit nanometer-thick outer layer of PEDOT onto a coaxial MnO_2/CNT composite [49]. Moreover, sonochemical processing was employed to wrap a water-soluble conducting polymer of PEDOT-PSS onto a binary composite of hierarchical MnO_2 nanospheres/CNTs, as shown in **Figure 6b** and **c** [50]. PEDOT-PSS not only served as a dispersant to stabilize the composite suspension, thus facilitating the electrode fabrication without use of binders, but also offered a good inter-particle connectivity between MnO_2 and CNTs.

5.4. Summary

Carbon nanostructures offer excellent substrates for enhancing the electrochemical performance of MnO_2. The carbon can be dispersed homogeneously into a solution, facilitating a uniform deposition of MnO_2 nanodeposits onto the carbon surface through a solution-based processing. The high surface area of carbon supports also provides a large number of anchoring sites for MnO_2 formation and enlarges the contact interfaces between MnO_2 and electrolyte. It is worthy to note that a porous carbon having well-ordered pore channels can permit fast electrolyte transport, meanwhile, a well-graphitic carbon with enhanced electrical conductivity can enable rapid charge collection/transfer under a high-rate operation. Conducting polymers can work in good synergy with MnO_2 to capture a maximum electrochemical harvesting from the large pseudocapacitances. A rational design strategy is necessary to control the MnO_2 dispersion into conducting polymers by tuning the synthesis conditions. For the *in situ* redox deposition, the mass loading of MnO_2 is extremely limited because an overoxidation of conducting polymers would destroy their π-conjugated structure, and accordingly, substantially lose their conductivities and electroactivities. A better synthesis alternative is to use MnO_2 nanostructures as reactive templates for polymerizing polymer monomers surrounding its outmost surfaces. The synchronous formation of MnO_2 and conducting polymers in the chemical co-precipitation method facilitates a molecular level interaction with each other, but the electrochemical performance depends on the morphologies of the nanocomposites. Interfacial synthesis is an optional approach to prepare a composite with mesoporous structure.

Ternary nanocomposites are capable of maximizing the desirable functions of each component. The synergistic contribution results in great capacitance enhancement, high-rate delivery and better cycling performance. Furthermore, it is important to bridge MnO_2 and carbon to electrolytes with a large interaction area by controlling the thickness of conducting polymer layers and their spatial distribution in the ternary structure.

6. Asymmetric supercapacitors based on MnO_2 nanostructures

6.1. Fabrication principles

To make an asymmetric supercapacitor that can operated stably in a wide potential range, the critical issue is to couple electrode materials with different high overpotentials for hydrogen or oxygen evolutions [51]. It is known that MnO_2 nanostructures normally work in the electrochemical potential window of 0.0–1.0 V (vs. SCE). Therefore, the coupled electrode materials should have a potential window in a range far beyond 0.0–1.0 V. Since the first report of MnO_2|| activated carbon asymmetric supercapacitor by Hong et al. in 2002 [52], carbon materials have attracted massive attention as coupled electrodes, because they can operate in a wide potential window from −1.0 to 0.1 V vs. SCE (Saturated Calomel Electrode) [17]. A wide operating potential window of 2.0 V is yielded in the assembled asymmetric supercapacitor in a Na_2SO_4 aqueous electrolyte. The asymmetric supercapacitor could even work in a working voltage as high as 2.3 V after elaborating the electrode materials (e.g., mass ratio) and cell assembly (e.g., oxygen expelling). Such a high-cell voltage is comparable to a commercially available symmetric supercapacitors using organic electrolyte (2.5 V) [51]. Consequently, the energy density of asymmetric supercapacitors is exceptionally high, reaching over 20 Wh/kg, almost one order of magnitude higher than that of the aqueous-based symmetric supercapacitor and also greater than that of organic-based SSCs (17–18 Wh/kg).

In addition to the carbon materials, pseudocapacitive materials, such as metal oxides and conducting polymers (e.g., PANi, PPy, and PEDOT), can be used as coupled electrodes for asymmetric supercapacitors. It is noted that the electrochemical oxidation/reduction potential is essentially associated with the work function of a metal oxide [53]. A large work function difference of two metal oxides provide an opportunity of enlarging operating voltage larger than the dissociation energy of an aqueous electrolyte because the water decomposition is kinetically limited by hydrogen and oxygen evolution reaction on the surface of metal oxides. Hence, a metal oxide with a largest work function difference from MnO_2 is more appropriate for asymmetric supercapacitor assembly to maximizing cell voltage in neutral aqueous electrolytes.

Moreover, another important technological issue is to balance the charge (Q) that stored at positive and negative electrodes. The charge stored by each electrode is linearly proportional to the specific capacitance (C), the potential window (ΔU), and the mass (m) of an electrode material, i.e., $Q = C \times \Delta U \times m$. In order to obtain charge balance, i.e., $Q_+ = Q_-$, the optimal mass ratio of the positive electrode (m_+) against negative electrode (m_-) can be estimated according to Eq. (8) [17]. It should be noted that the optimal mass ratio is responsible for an ideal capacitive behavior with a maximum cell voltage and a high-Coulombic efficiency.

$$\frac{m_+}{m_-} = \frac{C_- \times \Delta U_-}{C_+ \times \Delta U_+} \tag{8}$$

6.2. Asymmetric supercapacitor cells

Carbon materials are of particular interest as negative electrodes for asymmetric supercapacitors due to their characteristics of good EDLC, a high-overpotential for hydrogen evolution, excellent electrical conductivity for high power delivery, and good chemical inertness for long cycle life. In this regard, various carbon materials have been explored including graphene, CNTs, activated carbon nanofibers (ACNFs), carbon sphere (CS), ordered mesoporous carbon, etc., to couple MnO_2.

Graphene-based materials have been received intensive attention for asymmetric supercapacitors. First, graphene itself exhibits outstanding physiochemical and capacitive properties. Second, the ultrathin two-dimensional structure of graphene provides a large surface area for anchoring MnO_2 nanostructures [54, 55]. In addition, graphene-based materials can also be compacted into free-standing electrodes, which endow asymmetric supercapacitor with flexibility and lightweight. The energy and power values would be more fascinating at a device level taking the additives (binders and conductors) and the current collectors into account. **Figure 7** demonstrated porous hybrid thin-film structures of MnO_2/graphene and graphene/Ag, which are highly flexible, mechanically robust, and ultra-lightweight using the ordinal filtration assembly method [55]. The flexible device extended the operating potential window to 1.8 V and showed an energy density of 50.8 Wh/kg.

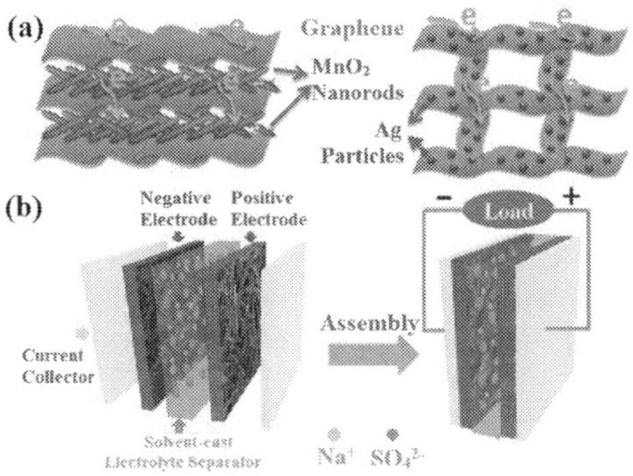

Figure 7. Illustrations of (a) MnO_2/graphene and graphene/Ag freestanding films and (b) the as-fabricated asymmetric supercapacitor devices.

CNTs and carbon nanofiber fabrics were demonstrated to serve excellent one-dimensional negative materials for energy storage devices due to their high electrical conductivity, good mechanical stability, and flexibility [56–60]. On one hand, MnO_2 nanostructures can be uniformly deposited surround the one-dimensional carbons and the resulting nanocomposites showed pronounced electrochemical properties. On the other hand, the porous one-dimensional carbons possess good electric double-layer capacitance, thus becoming highly promising negative candidates for asymmetric supercapacitors. The asymmetric supercapacitors assembled by one-dimensional carbons as the negative electrodes and MnO_2/carbon nanocomposites as the positive electrode can be cycled reversibly in a high-voltage region of 0–2.0 V and exhibited a superior energy density of 30–35 Wh/kg in a neutral aqueous Na_2SO_4 electrolyte. In particular, the asymmetric supercapacitors based on electrospun carbon nanofibers are rather promising to meet the ever-increasing demands of high energy/power densities because of the scalability, easy-fabrication, and low-cost of the electrospinning technique [56].

In addition to the carbon materials, metal oxides and conducting polymers, such as SnO_2, MoO_3 PANi, PPy, and PEDOT, were also explored as negative materials to couple MnO_2 nanostructures [51, 61]. MoO_3 would be the most favorable coupled pair of MnO_2 because of their largest work function difference [53]. To this end, MoO_3/graphene composite and MnO_2/graphene nanocomposite were prepared to serve as the negative electrode and the positive electrode for the fabrication of an asymmetric supercapacitor. The as-assembled device can operated stably in a wide voltage window of 2.0 V, and more significantly, exhibited a high-specific capacitance of 307 F/g and a high-energy density of 42.6 Wh/kg.

7. Summaries and perspectives

MnO_2-based electrode materials hold a particular prospect for future supercapacitor applications although there are still some obstacles that need to be conquered. To better utilize the electrochemical performance of MnO_2, it is necessary to create more electrochemically active sites or reduce the ion/electron transport distance by modifying surface chemistry and structure of MnO_2. One of the most effective approaches is incorporating nanoscaled MnO_2 into a highly porous and electronically conductive framework, such as carbon scaffolds and conducting polymers, to form a hybrid material. Benefiting from the synergistic contribution from each component in the composite, the consistent progress in the research and development of MnO_2-based nanocomposites have achieved many breakthroughs in terms of charge storage mechanism, smart design strategies, and technological innovations. It should be noted that more advanced characterization tools and new methodologies as well systematic studies are still required to deepen the fundamental understanding of material chemistry and electrode/electrolyte interface. The energy and power densities of supercapacitors can be significantly improved by one magnitude through a simple asymmetric configuration method. Although many limiting factors, such as optimal coupled electrode, high-cost issue, complex fabrication process, moderate cycle life, etc., are required to be conquered urgently, it is believed that the ongoing research would bring MnO_2-based asymmetric supercapacitors to an acceptable level for practical applications.

Acknowledgements

The author acknowledge the financial supports of this work by the National Natural Science Foundation of China (51402236), the Natural Science Foundation of Shaanxi Province (2015JM5180), the Research Fund of the State Key Laboratory of Solidification Processing (NWPU), China (Grant No.: 123-QZ-2015), and the Key Laboratory of New Ceramic and Fine Processing (Tsinghua University, KF201607).

Author details

Jian-Gan Wang

Address all correspondence to: wangjiangan@nwpu.edu.cn

State Key Laboratory of Solidification Processing, Center for Nano Energy Materials, School of Materials Science and Engineering, Northwestern Polytechnical University, Xi'an, China

References

[1] Salanne M, Rotenberg B, Naoi K, Kaneko K, Taberna PL, Grey CP, Dunn B, Simon P. Efficient storage mechanisms for building better supercapacitors. Nature Energy. 2016; 1: 16070. doi:10.1038/NENERGY.2016.70

[2] Lee HY, Goodenough JB. Supercapacitor behavior with KCl electrolyte. Journal of Solid State Chemistry. 1999; 144: 220–223. doi:10.1006/jssc.1998.8128

[3] Toupin M, Brousse T, Bélanger D. Charge storage mechanism of MnO_2 electrode used in aqueous electrochemical capacitor. Chemistry of Materials. 2004; 16: 3184–3190. doi: 10.1021/cm049649j

[4] Wei W, Cui X, Chen W, Ivey DG. Manganese oxide-based materials as electrochemical supercapacitor electrodes. Chemical Society Reviews. 2011; 40: 1697–1721. doi: 10.1039/C0CS00127A

[5] Xu C, Kang F, Li B, Du H. Recent progress on manganese dioxide based supercapacitors. Journal of Materials Research. 2010; 25: 1421–1432. doi:10.1557/JMR.2010.0211

[6] Zhai T, Lu X, Wang F, Xia H, Tong Y. MnO_2 nanomaterials for flexible supercapacitors: performance enhancement via intrinsically and extrinsically modification. Nanoscale Horizons. 2016; 1: 109–124. doi:10.1039/c1035nh00048c.

[7] Ghodbane O, Pascal JL, Favier F. Microstructural effects on charge-storage properties in MnO_2-based electrochemical supercapacitors. ACS Applied Materials and Interfaces. 2009; 1: 1130–1139. doi:10.1021/am900094e

[8] Ghodbane O, Pascal JL, Fraisse B, Favier F. Structural *in situ* study of the thermal behavior of manganese dioxide materials: toward selected electrode materials for supercapacitors. ACS Applied Materials and Interfaces. 2010; 2: 3493–3505. doi: 10.1021/am100669k

[9] Kuo SL, Wu NL. Investigation of pseudocapacitive charge-storage reaction of $MnO_2°$ nH_2O supercapacitors in aqueous electrolytes. Journal of The Electrochemical Society. 2006; 153: A1317–A1324. doi:10.1149/1.2197667

[10] Ghaemi M, Ataherian F, Zolfaghari A, Jafari S. Charge storage mechanism of sonochemically prepared MnO_2 as supercapacitor electrode: effects of physisorbed water and proton conduction. Electrochimica Acta. 2008; 53: 4607–4614. doi:10.1016/j.electacta.2007.12.040

[11] Brousse T, Toupin M, Dugas R, Athouël L, Crosnier O, Bélanger D. Crystalline MnO_2 as possible alternatives to amorphous compounds in electrochemical supercapacitors. Journal of The Electrochemical Society. 2006; 153: A2171–A2180. doi:10.1149/1.2352197

[12] Devaraj S, Munichandraiah N. Effect of crystallographic structure of MnO_2 on its electrochemical capacitance properties. The Journal of Physical Chemistry C. 2008; 112: 4406–4417. doi:10.1021/jp7108785

[13] Chang JK, Lee MT, Tsai WT, Deng MJ, Sun IW. X-ray photoelectron spectroscopy and *in situ* X-ray absorption spectroscopy studies on reversible insertion/desertion of dicyanamide anions into/from manganese oxide in ionic liquid. Chemistry of Materials. 2009; 21: 2688–2695. doi:10.1021/cm9000569

[14] Song MK, Cheng S, Chen H, Qin W, Nam KW, Xu S, Yang XQ, Bongiorno A, Lee J, Bai J. Anomalous pseudocapacitive behavior of a nanostructured, mixed-valent manganese oxide film for electrical energy storage. Nano Letters. 2012; 12: 3483–3490. doi: 10.1021/nl300984y

[15] Wang JG, Yang Y, Huang ZH, Kang F. Shape-controlled synthesis of hierarchical hollow urchin-shape α-MnO_2 nanostructures and their electrochemical properties. Materials Chemistry and Physics. 2013; 140: 643–650. doi:10.1016/j.matchemphys.2013.04.018

[16] Wang JG, Yang Y, Huang ZH, Kang F. Interfacial synthesis of mesoporous MnO_2/polyaniline hollow spheres and their application in electrochemical capacitors. Journal of Power Sources. 2012; 204: 236–243. doi:10.1016/j.jpowsour.2011.12.057

[17] Wang JG, Kang F, Wei B. Engineerging of MnO_2-based nanocomposites for high performance supercapacitors. Progress in Materials Science. 2015; 74: 51–124. doi: 10.1016/j.pmatsci.2015.04.003

[18] Zhang H, Cao G, Wang Z, Yang Y, Shi Z, Gu Z. Growth of manganese oxide nanoflowers on vertically-aligned carbon nanotube arrays for high-rate electrochemical capacitive energy storage. Nano Letters. 2008; 8: 2664–2668. doi:10.1021/nl800925j

[19] Liu R, Lee SB. MnO_2/poly (3, 4-ethylenedioxythiophene) coaxial nanowires by one-step coelectrodeposition for electrochemical energy storage. Journal of the American Chemical Society. 2008; 130: 2942–2943. doi:10.1021/ja7112382

[20] Liu J, Essner J, Li J. Hybrid supercapacitor based on coaxially coated manganese oxide on vertically aligned carbon nanofiber arrays. Chemistry of Materials. 2010; 22: 5022–5030. doi:10.1021/cm101591p

[21] Sharma RK, Zhai L. Multiwall carbon nanotube supported poly (3, 4-ethylenedioxythiophene)/manganese oxide nano-composite electrode for super-capacitors. Electrochimica Acta. 2009; 54: 7148–7155. doi:10.1016/j.electacta.2009.07.048

[22] Zhu J, He J. Facile synthesis of graphene-wrapped honeycomb MnO_2 nanospheres and their application in supercapacitors. ACS Applied Materials and Interfaces. 2012; 4: 1770–1776. doi:10.1021/am3000165

[23] Wang JG, Yang Y, Huang ZH, Kang F. Effect of Fe^{3+} on the synthesis and electrochemical performance of nanostructured MnO_2. Materials Chemistry and Physics. 2012; 133: 437–444. doi:10.1016/j.matchemphys.2012.01.056

[24] Ragupathy P, Park DH, Campet G, Vasan HN, Hwang SJ, Choy JH, Munichandraiah N. Remarkable capacity retention of nanostructured manganese oxide upon cycling as an electrode material for supercapacitor. The Journal of Physical Chemistry C. 2009; 113: 6303–6309. doi:10.1021/jp811407q

[25] Hu CC, Wang CC. Nanostructures and capacitive characteristics of hydrous manganese oxide prepared by electrochemical deposition. Journal of The Electrochemical Society. 2003; 150: A1079–A1084. doi:10.1149/1.1587725

[26] Wei W, Cui X, Mao X, Chen W, Ivey DG. Morphology evolution in anodically electrodeposited manganese oxide nanostructures for electrochemical supercapacitor applications—effect of supersaturation ratio. Electrochimica Acta. 2011; 56: 1619–1628. doi:10.1016/j.electacta.2010.10.044

[27] Xu CL, Bao SJ, Kong LB, Li H, Li HL. Highly ordered MnO_2 nanowire array thin films on Ti/Si substrate as an electrode for electrochemical capacitor. Journal of Solid State Chemisitry. 2006; 179: 1351–1355. doi:10.1016/j.jssc.2006.01.058

[28] Wei J, Zhitomirsky I. Electrosynthesis of manganese oxide films. Surface Engineering. 2008; 24: 40–46. doi:10.1179/174329408X271363

[29] Yuan C, Gao B, Shen L, Yang S, Hao L, Lu X, Zhang F, Zhang L, Zhang X. Hierarchically structured carbon-based composites: design, synthesis and their application in electrochemical capacitors. Nanoscale. 2011; 3: 529–545. doi:10.1039/C0NR00423E

[30] Zhang LL, Zhao XS. Carbon-based materials as supercapacitor electrodes. Chemical Society Reviews. 2009; 38: 2520–2531. doi:10.1039/B813846J

[31] Wang JG, Yang Y, Huang ZH, Kang F. Coaxial carbon nanofibers/MnO_2 nanocomposites as freestanding electrodes for high-performance electrochemical capacitors. Electrochimica Acta. 2011; 56: 9240–9247. doi:10.1016/j.electacta.2011.07.140

[32] Wang JG, Yang Y, Huang ZH, Kang F. Synthesis and electrochemical performance of MnO_2/CNTs–embedded carbon nanofibers nanocomposites for supercapacitors. Electrochimica Acta. 2012; 75:213–219. doi:10.1016/j.electacta.2012.04.088

[33] Xia H, Wang Y, Lin J, Lu L. Hydrothermal synthesis of MnO_2/CNT nanocomposite with a CNT core/porous MnO_2 sheath hierarchy architecture for supercapacitors. Nanoscale Research Letters. 2012; 7: 1–10. doi:10.1186/1556-276X-7-33

[34] Chen W, Fan Z, Gu L, Bao X, Wang C. Enhanced capacitance of manganese oxide via confinement inside carbon nanotubes. Chemical Communications. 2010; 46: 3905–3907. doi:10.1039/c000517g

[35] Shao Y, El-Kady MF, Wang LJ, Zhang Q, Li Y, Wang H, Mousavi MF, Kaner RB. Graphene-based materials for flexible supercapacitors. Chemical Society Reviews. 2015; 44: 3639–3665. doi:10.1039/c4cs00316k

[36] Yan J, Fan Z, Wei T, Qian W, Zhang M, Wei F. Fast and reversible surface redox reaction of graphene–MnO_2 composites as supercapacitor electrodes. Carbon. 2010; 48: 3825–3833. doi:10.1016/j.carbon.2010.06.047

[37] Chen S, Zhu J, Wu X, Han Q, Wang X. Graphene oxide–MnO_2 nanocomposites for supercapacitors. ACS Nano. 2010; 4: 2822–2830. doi:10.1021/nn901311t

[38] Lei Z, Shi F, Lu L. Incorporation of MnO_2-coated carbon nanotubes between graphene sheets as supercapacitor electrode. ACS Applied Materials and Interfaces. 2012; 4: 1058–1064. doi:10.1021/am2016848

[39] Jiang H, Dai Y, Hu Y, Chen W, Li C. Nanostructured ternary nanocomposite of rGO/CNTs/MnO_2 for high-rate supercapacitors. ACS Sustainable Chemistry and Engineering. 2013; 2: 70–74. doi:10.1021/sc400313y

[40] Jiang H, Ma J, Li C. Mesoporous carbon incorporated metal oxide nanomaterials as supercapacitor electrodes. Advanced Materials. 2012; 24: 4197–4202. doi:10.1002/adma.201104942

[41] Dong X, Shen W, Gu J, Xiong L, Zhu Y, Li H, Shi J. MnO_2-embedded-in-mesoporous-carbon-wall structure for use as electrochemical capacitors. The Journal of Physical Chemistry B. 2006; 110: 6015–6019. doi:10.1021/jp056754n

[42] Jiang H, Yang L, Li C, Yan C, Lee PS, Ma J. High-rate electrochemical capacitors from highly graphitic carbon–tipped manganese oxide/mesoporous carbon/manganese oxide hybrid nanowires. Energy and Environmental Science. 2011; 4: 1813–1819. doi:10.1039/c1ee01032h

[43] Han J, Li L, Fang P, Guo R. Ultrathin MnO_2 nanorods on conducting polymer nanofibers as a new class of hierarchical nanostructures for high-performance supercapacitors. The Journal of Physical Chemistry C. 2012; 116: 15900–15907. doi:10.1021/jp303324x

[44] Jafri RI, Mishra AK, Ramaprabhu S. Polyaniline–MnO_2 nanotube hybrid nanocomposite as supercapacitor electrode material in acidic electrolyte. Journal of Materials Chemistry. 2011; 21: 17601–17605. doi:10.1039/c1jm13191e

[45] Zang J, Li X. *In situ* synthesis of ultrafine β-MnO_2/polypyrrole nanorod composites for high-performance supercapacitors. Journal of Materials Chemistry. 2011; 21: 10965–10969. doi:10.1039/c1jm11491c

[46] Wang JG, Yang Y, Huang ZH, Kang F. MnO_2/polypyrrole nanotubular composites: reactive template synthesis, characterization and application as superior electrode materials for high-performance supercapacitors. Electrochimica Acta. 2014; 130: 642–649. doi:10.1016/j.electacta.2014.03.082

[47] Wang JG, Yang Y, Huang ZH, Kang F. Facile synthesis of hierarchical conducting polypyrrole nanostructures via a reactive template of MnO_2 and their application in supercapacitors. RSC Advances. 2014; 4: 199–202. doi:10.1039/c3ra45824e

[48] Lu Q, Zhou Y. Synthesis of mesoporous polythiophene/MnO_2 nanocomposite and its enhanced pseudocapacitive properties. Journal of Power Sources. 2011; 196: 4088–4094. doi:10.1016/j.jpowsour.2010.12.059

[49] Yoon SB, Kim KB. Effect of poly (3, 4-ethylenedioxythiophene)(PEDOT) on the pseudocapacitive properties of manganese oxide (MnO_2) in the PEDOT/MnO_2/multiwall carbon nanotube (MWNT) composite. Electrochimica Acta. 2013: 106: 135–142. doi:10.1016/j.electacta.2013.05.058

[50] Hou Y, Cheng Y, Hobson T, Liu J. Design and synthesis of hierarchical MnO_2 nanospheres/carbon nanotubes/conducting polymer ternary composite for high perform-

ance electrochemical electrodes. Nano Letters. 2010; 10: 2727–2733. doi:10.1021/nl101723g

[51] Zhang J, Zhao XS. On the configuration of supercapacitors for maximizing electrochemical performance. ChemSusChem. 2012; 5: 818–841. doi:10.1002/cssc.201100571

[52] Hong MS, Lee SH, Kim SW. Use of KCl aqueous electrolyte for 2 V manganese oxide/activated carbon hybrid capacitor. Electrochemical Soild-State Letters. 2002; 5: A227–A230. doi:10.1149/1.1506463

[53] Chang J, Jin M, Yao F, Kim TH, Le VT, Yue H, Gunes F, Li B, Ghosh A, Xie S, Lee YH. Asymmetric supercapacitors based on graphene/MnO_2 nanospheres and graphene/MoO_3 nanosheets with high energy density. Advanced Functional Materials. 2013; 23: 5074–5083. doi:10.1002/adfm201301851

[54] Wu ZS, RenW, Wang DW, Li F, Liu B, Cheng HM. High-energy MnO_2 nanowire/graphene and graphene asymmetric electrochemical capacitors. ACS Nano. 2010; 4: 5835–5842. doi:10.1021/nn101754k

[55] Shao Y, Wang H, Zhang Q, Li Y. High-performance flexible asymmetric supercapacitors based on 3D porous graphene/MnO_2 nanorod and graphene/Ag hybrid thin-film electrodes. Journal of Materials Chemistry C. 2013; 1: 1245–1251. doi:10.1039/c2tc00235c

[56] Wang JG, Yang Y, Huang ZH, Kang F. A high-performance asymmetric supercapacitor based on carbon and carbon-MnO_2 nanofiber electrodes. Carbon. 2013; 61:190–199. doi:10.1016/j.carbon.2013.04.084

[57] Yu G, Hu L, Vosgueritchian M, Wang H, Xie X, McDonough JR, Cui X, Cui Y, Bao Z. Solution-processed graphene/MnO_2 nanostructured textiles for high-performance electrochemical capacitors. Nano Letters. 2011; 11: 2905–2911. doi:10.1021/nl2013828

[58] Wang JG, Yang Y, Huang ZH, Kang F. Incorporation of nanostructured manganese dioxide into carbon nanofibers and its electrochemical performance. Materials Letters. 2012; 72: 18–21. doi:10.1016/j.matlet.2011.12.071

[59] Wang JG, Yang Y, Huang ZH, Kang F. Rational synthesis of MnO_2/conducting polypyrrole@carbon nanofiber triaxial nano-cables for high-performance supercapacitors. Journal of Materials Chemistry. 2012; 22: 16943–16949. doi:10.1039/c2jm33364c

[60] Wang JG, Yang Y, Huang ZH, Kang F. Effect of temperature on the pseudo-capacitive behavior of freestanding MnO_2@ carbon nanofibers composites electrodes in mild electrolyte. Journal of Power Sources. 2013; 224: 86–92. doi:0.1016/j.jpowsour.2012.09.075

[61] Ng KC, Zhang S, Chen GZ. An asymmetrical supercapacitor based on CNTs/SnO_2 and CNTs/MnO_2 nanocomposites working at 1.7 V in aqueous electrolyte. ECS Transactions. 2008; 16: 153–162. doi:10.1149/1.2985638

Chapter 4

Nanostructured Metal Oxides-Based Electrode in Supercapacitor Applications

Zhenjun Qi, Shihao Huang, Adnan Younis,
Dewei Chu and Sean Li

Additional information is available at the end of the chapter

http://dx.doi.org/10.5772/65155

Abstract

To overcome the obstacle of low energy density, one of the most intensive approaches is the development of new materials for supercapacitor electrodes. Most explored materials today are carbon particle materials, which have high surface areas for charge storage. But in spite of these large specific surface areas, the charges physically stored on the carbon particles in porous electrode layers are unfortunately limited. Regarding advanced supercapacitor electrodes, metal oxides are considered the most promising material for the next generation of supercapacitors owing to their unique physical and chemical properties. In this chapter, the rational design and fabrication of metal oxide nanostructures for supercapacitor applications are addressed.

Keywords: supercapacitor, metal oxide, nanostructure, nanocomposite, graphene

1. Introduction

In recent years, supercapacitors (or ultracapacitors) have attracted significant attention as a versatile solution to meet the increasing demands of energy storage because of their fast power energy delivery, long lifecycle, high power density and reasonably high energy density which are able to fill in the gap between the batteries and the conventional capacitors [1–3].

The comparison of specific power and specific energy for different electrical energy storage devices is shown in the Ragone plot (**Figure 1**) [4]. The data illustrate that supercapacitors are able to store more energy than conventional capacitors and deliver more power than batteries. Owing to the different energy storage mechanism from conventional capacitors, the specific energy of supercapacitors can be thousands of times higher than it of conventional capacitors by forming an electric double layer at the interface of the electrode and electrolyte to store energy [5]. Thus, the high surface area of the electrode can be adequately utilized to collect amounts of positively and negatively charged ions from electrolyte when storing energy.

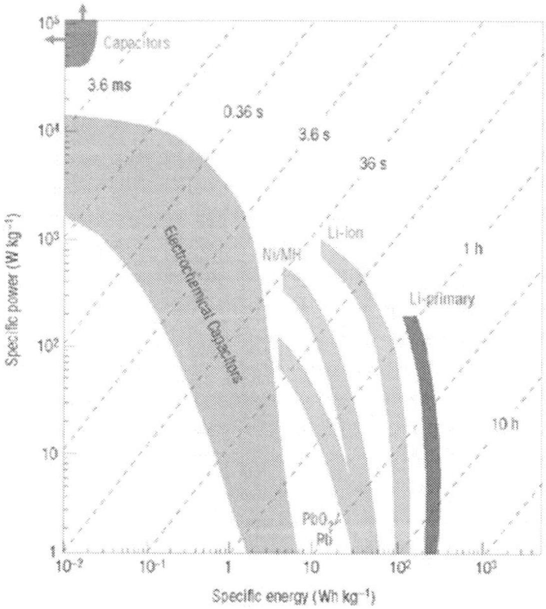

Figure 1. Specific power and energy for various electrical energy storage devices show in Ragone plot [4].

On the other hand, as is well-known, rechargeable batteries mainly depend on chemical reactions to charge and discharge which significantly restrict their lifetime [6]. Compared to batteries, the energy storage process of supercapacitors is based on electrostatic storage in the electrical double layer and reversible faradaic redox reactions by means of electron charge transfer on the surface of electrode. Thus, supercapacitors are expected to have a capability of faster charge–discharge under high current and a longer cycle life than batteries because no or negligibly few chemical reactions are involved [7].

Even though major progress has been yielded in the theoretical and practical research and development of supercapacitors, few disadvantages of supercapacitors, including low energy density and high production cost, have been identified as major challenges for the furtherance of supercapacitors technologies [8].

To overcome the obstacle of low energy density, one of the most intensive approaches is the development of new materials for supercapacitor electrodes. Most explored materials today are carbon particle-based materials, which have high surface areas for charge storage [3]. But

http://dx.doi.org/10.5772/65155

in spite of these large specific surface areas, the charges physically stored on the carbon particles in porous electrode layers unfortunately limiting their electrochemical properties. Supercapacitors of this kind, called electrical double-layer supercapacitors (EDLS), have a limited specific capacitance and relative low energy density [9]. Supercapacitors with electrochemically active materials (polymers and metal oxides) as electrodes involving fast and reversible faradaic reactions on electrodes are called faradaic supercapacitors (FS). It has been demonstrated that faradaic or hybrid double-layer supercapacitors can yield much higher specific capacitance and energy density. Thus, regarding advanced supercapacitor materials, metal oxides are considered the most promising materials for the next generation of supercapacitors [10].

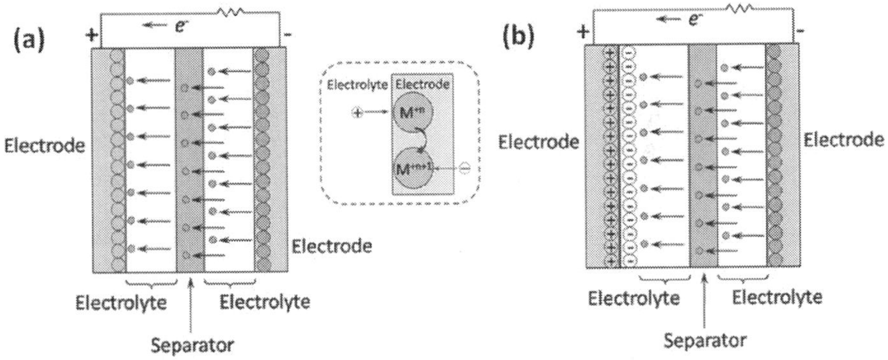

Figure 2. Mechanisms of (a) pseudocapacitance and (b) hybrid capacitance [13].

The capacitance of a metal oxide-based supercapacitor is determined by two storage principles, one is double-layer capacitance and another one is pseudocapacitance. The mechanism of double-layer capacitance is shown in **Figure 2a**, when capacitor charged, electrostatic storage achieved by separation of charge in a double layer at the interface between the surface of a conductive electrode and an electrolyte. As a result, mirror image of charge distribution of ions in opposite polarity, called double-layer, is formed. When capacitor discharged, ions return and distribute randomly in the electrolyte. For pseudocapacitance, it stores electrical energy electrochemically by means of reversible faradaic redox reactions on the surface of suitable electrode materials in an electrochemical capacitor with an electric double-layer [11]. It can be seen in **Figure 2b**, pseudocapacitance is accompanied with an electron charge-transfer between electrolyte and electrode coming from a de-solvated and adsorbed ion whereby only one electron per charge unit is participating. This faradaic charge transfer originates via a very fast sequence of reversible redox, electrosorption or intercalation processes. The adsorbed ion has no chemical reaction with the atoms of the electrode. No chemical bonds arise, and only a charge-transfer takes place [12]. Even though both double-layer capacitance and pseudocapacitance contribute indivisibly to the total capacitance of metal oxide supercapacitors, the latter can be 10–100 times higher than the former.

In general, metal oxide-based supercapacitors are able to possess higher specific capacitance and energy density than carbon materials and conducting polymer materials [14]. A series of metal oxides with high theoretical performances has been studied, such as RuO_2, MnO_2, NiO, Co_3O_4, V_2O_5, CuO and Fe_3O_4. **Table 1** [15] lists the theoretical capacitance of some typical metal oxides as well as the charge storage reactions. However, the practical supercpacitive properties of these metal oxides are far behind their theoretical values due to their low conductivities, poor long-term stability, low surface areas and porosity.

Oxide	Electrolyte	Charge storage reaction	Theoretical capacitance (F g^{-1})	Conductivity (S cm^{-1})
MnO_2	Na_2SO_4	$MnO_2 + M^+ + e^- = MMnO_2$ (M could be H^+, Li^+, Na^+, K^+)	1380	10^{-5} to 10^{-6}
V_2O_5	NaCl, Na_2SO_4	$V_2O_5 + 4M^+ + 4e^- = M_2V_2O_5$ (M could be H^+, Li^+, Na^+, K^+)	2120	10^{-4} to 10^{-2}
NiO	KOH, NaOH	$NiO + OH^- = NiOOH + e^-$	2584	0.01 to 0.32
Co_3O_4	KOH, NaOH	$Co_3O_4 + OH^- + H_2O = 3CoOOH + e^-$ $CoOOH + OH^- = CoO_2 + H_2O + e^-$	3560	10^{-4} to 10^{-2}

Table 1. Pseudocapacitance and conductivity of selected metal oxides.

In this chapter, several important factors affecting the electrochemical properties of metal oxide-based electrodes are discussed firstly. Then various methods to fabricate nanostructured metal oxide electrode are summarized. Finally, advanced metal oxide-carbon composite electrodes are further described.

2. Factors affecting the performance of metal oxide-based electrodes

2.1. Crystallinity

The degree of crystallinity is one of pivotal factors affecting the pseudocapacitance of metal oxide materials. In general, an amorphous structure exhibits superior electrochemical performance than a well-crystallized structure due to the former can make the fast, continuous and reversible faradaic reaction take place not only on the surface but also inside of metal oxide particles [16]. This is because the amorphous structure with a highly porous morphology is benefit for ion accessibility and cation diffusion. In addition, these porous structures also result in a higher specific surface area which can support more redox reactions to enhance the specific capacitance. Nevertheless, it is well-known that the poorly crystallized metal oxide simultaneously leads to a lower electrical conductivity limiting its pseudocapacitance. Thus, it is necessary and a great challenge to explore the appropriate crystallinity with optimal conductivity and ionic transport.

One method is to improve the electric conductivity of the amorphous metal oxide. It has been reported that annealing can significantly affect the electrical conductivity. For example, in [17], MnOx annealed at 200°C exhibited a higher specific capacitance at high scan rate than those without treated. Another effective approach is to optimize the structure of crystallized metal oxide in order to provide appropriate tunnels for the intercalation of cations.

2.2. Crystal structure

The crystal structure has a significant influence on the pseudocapacitance of metal oxide because it plays a crucial role in determining the cations intercalation. For instance, crystallized manganese oxide has different crystalline structures, including α-, β-, γ-, δ- and λ-MnO_2 shown in **Figure 3** [18]. It can be seen thatα-MnO_2 forms 1D (2 × 2) and (1 × 1) tunnels; β-MnO_2 forms a 1D (1 × 1) tunnel; γ-MnO_2 is consist of 1D (1 × 2) and (1 × 1) tunnels; δ-MnO_2 is a 2D layered structure; and λ-MnO_2 is a three-dimensional (3D) spinel structure, respectively [11, 19]. It is reported by Brousse et al. [20] that α-MnO_2 with a large tunnel size exhibited a relatively high specific capacitance of 110 F g^{-1} due to K^+ cations could easily insert the tunnels. On the contrary, β-MnO_2 with a narrow tunnel size which is smaller than K^+ cations inhibited the diffusion process and leaded to a low specific capaci-

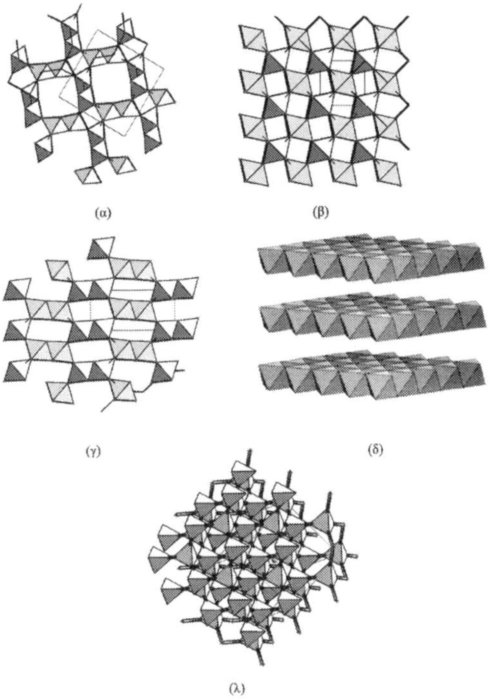

Figure 3. Crystal structures of α-, β-, γ-, δ- and λ-MnO_2 [18].

tance value of only 110 μF cm^{-2}. The result indicated that the limited performance was manly obtained on the surface of manganese oxide. Furthermore, the 2D δ-MnO_2 with an interlayer separation of around 0.7 nm also obtained good capacitance value of 236 F g^{-1} which could be contributed to the sufficiently large layer space for a high rate insertion/extraction of K^+ cations.

Thus, the rational selection of crystal structures can effectively accelerate the charge storage process of metal oxide and improve its electrochemical properties.

2.3. Specific surface area

The pseuocapacitance of metal oxide depends on redox reactions which mainly take place on the surface area. Thus, the specific surface area is one of the most important factors for metal oxide-based supercapacitor applications. In general, the higher specific surface area can result in the higher the specific capacitance due to more active sites are capable of providing multiple redox reactions [21].

Obviously, to explore the higher specific surface area is an effective approach to achieve better capacitive performance of metal oxide. Up to now, amounts of attempts have been made on metal oxides including decreasing the size of their particles, optimizing their morphologies and combining them with carbon materials which have high specific surface areas.

2.4. Morphology

The morphology of metal oxide is a crucial factor which closely relates to the specific surface area, the diffusion pathway, surface to volume ratios and therefore the supercapacitive performance. Thus, considerable efforts have been focused on various metal oxides with different morphologies such as nanowires, nanorods, nanotubes, nanoflowers, hollow spheres, nanopillar array and porous thin films.

Several morphologies of metal oxide and their supercapacitive performance are discussed in this part. The first morphology is one-dimensional nanostructured metal oxides which generally enhance the specific capacitance through offering short diffusion path lengths for both ions and electrons as well as a large specific surface area. Gao et al. [22] successfully synthesized Co_3O_4 nanowires on nickel foam via template-free method shown in **Figure 4a**. The nanowires, with diameters around 250 nm and the lengths up to around 15 μm, displayed a maximum specific capacitance of 746 F g^{-1} at a current density of 5 mA cm^{-2}. In addition, Lu et al. [23] reported a slim (<20 nm) NiO nanorod structure (**Figure 4b**) had an ultrahigh specific capacitance of 2018 F g^{-1} (80% of the theoretical value) at a current density of 2.27 A g^{-1} and high power density of 1536 F g^{-1} at 22.7 A g^{-1}. Generally, the diameter plays an important role in one-dimensional nanostructure. The smaller diameter can offer larger specific surface area and more active sites leading to a better specific capacitance. It is also reported that the porous nanotube structure of MnO_2 could not only enhance the specific capacitance, but also improve the stability of electrode due to accommodating large volume charges during the charge-discharge cycle [24]. The second metal oxide structure should be noted is hollow spheres. Various metal oxides with hollow sphere morphologies have been successfully synthesized recently to pursue a high loosely mesoporous structure, large specific surface area and fast ion/electron transfer. For example, Yan et al. [25] fabricated hierarchically porous NiO hollow spheres composed of nanoflakes shown in **Figure 4c** with thicknesses of ~10 nm via a powerful chemical bath deposition method. The specific capacitance of NiO hollow spheres can remain 346 F g^{-1} at 1 A g^{-1} after 2000 cycles indicating an excellent supercapacitive performance. Another example is [26] that Co_3O_4 hollow spheres prepared by a facile carbonaceous microsphere templated synthesis as shown in **Figure 4d**. The as-obtained Co_3O_4 hollow spheres are composed of nanoparticles and possess a high surface area of 60 m^2 g^{-1} owing to their mesoporous structure. Such a unique hollow-sphere architecture can greatly contributed to the comparatively high capacitance and excellent cycling stability. The third type of metal oxide morphology is three-dimensional porous structure. It can be seen in **Figure 4e** that 3D highly ordered nanoporous CuO with interconnected bimodal nanopores were fabricated by Moosavifard and coworkers [27]. This morphology offered a large specific surface area of 149 m^2 g^{-1} displayed an excellent specific capacitance of 431 F g^{-1} at 3.5 mA cm^{-2} due to the 3D porous structure providing facilitated ion transport, short ion and electron diffusion pathways and more active sites for electrochemical reactions. Moreover, 3D-nanonet hollow structured Co_3O_4,

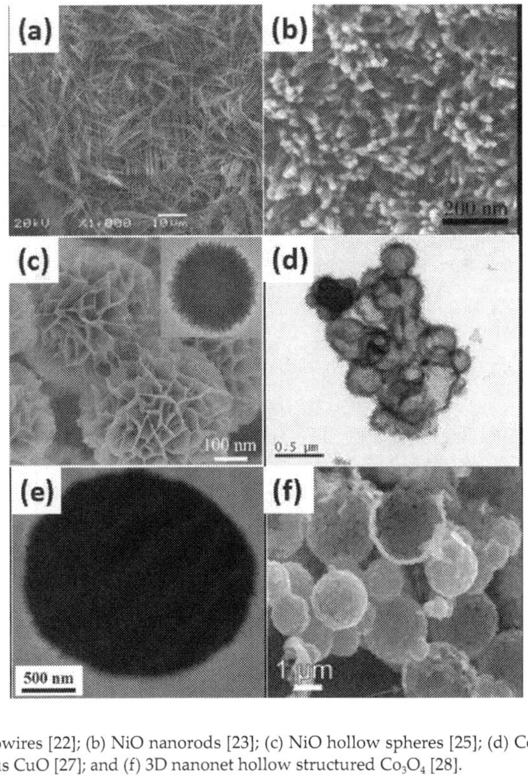

Figure 4. (a) Co_3O_4 nanowires [22]; (b) NiO nanorods [23]; (c) NiO hollow spheres [25]; (d) Co_3O_4 hollow spheres [26]; (e) 3D highly nanoporous CuO [27]; and (f) 3D nanonet hollow structured Co_3O_4 [28].

shown in **Figure 4f**, exhibited a maximum specific capacitance of 820 F g^{-1} at a scan rate of 5 mV s^{-1} and remained 90.2% of its initial capacitance after 1000 cycles [28]. The above results have proven that three-dimensional porous structured metal oxide is promising for supercapacitor applications.

As mentioned above, the electrochemical properties of metal oxide-based supercapacitors are largely affected by the morphologies. The rational design of morphologies with high specific surface area and porous nanostructures is necessary for the development of metal oxide-based electrodes.

2.5. Conductivity

It is well-known that the electronic conductivity of electrode materials is a vital factor to affect their high performance in supercapacitor applications. Unfortunately, the electronic conductivity of metal oxide materials is generally poor which largely limits ion and electron transfer. When the charge/discharge rate increases, the low conductivity would result in the localized charge/discharge process in a limited volume near the current collector, leading to low specific capacitance and low rate capability. For example, due to the low electronic conductivity ($\sim 10^{-5}$ to 10^{-6} S cm^{-1}), the realistic specific capacitance of manganese oxide is usually up to 350 F g^{-1} far behind its theoretic value of 1370 F g^{-1} [15, 29]. The same condition also takes place on other metal oxide materials such as NiO, Co_3O_4 and V_2O_5 [30, 31].

Thus, it is urgent to improve the conductivity of metal oxide in order to an ideal supercapacitive performance. One effective approach is doping metal elements into metal oxide and compositing metal oxide with high conductive carbon materials and conducting polymer.

2.6. Mass loading

The quantity of active materials loading on substrates can affect the specific capacitance, power and energy performance of metal oxide electrodes. On one hand, the large mass loading can cause longer transport paths for the diffusion of protons and an increase of thickness of metal oxide thin films resulting in lower electrical conductivity, limited access of electrolyte ions and higher series resistance. As a result, only partial active material on the surface of electrode film takes part in the charge storage leading to a lower specific capacitance of metal oxide. It is reported by Yang et al. [32] the specific capacitance of MnO_2 thin films decreased from 203 to 155 F g^{-1} when the mass loading increased from 6 to 18 mg cm^{-2}. On the other hand, a high mass loading is needed for high power and energy density, which thus makes the application of a light and durable supercapacitor possible [11]. Therefore, it is still a challenge for metal oxide to achieve both high mass loading and excellent specific capacitance.

3. Typical synthesis methods of metal oxide nanostructures

Nanostructured metal oxide materials have been intensively investigated due to their superior supercapacitive performance. The factors mentioned above are all related to the metal oxide fabrication processes and parameters. The main synthesis techniques exploited include hydrothermal, electrodeposition, sol–gel, microwave assisted as well as template assisted methods.

3.1. Hydrothermal method

Hydrothermal synthesis is well-known as one of the most outstanding approaches to prepare nanoparticles due to a serious of advantages such as fine powder (nanoscale), high purity, good dispersion, uniform, narrow distribution, without agglomeration, good crystal form and shape controllability. In a hydrothermal process, crystal grows by chemical reactions taking place at high temperature and pressure conditions in a sealed pressure vessel with water as solvent. Under hydrothermal conditions, water can act as a chemical component and participate in the reactions. Moreover, the solvent is not only a mineralizing agent but also a pressure medium. By the control of physical and chemical factors, the formation and modification of nanostructured metal oxide can be achieved. Up to now, hydrothermal method has been successfully used to synthesize metal oxide with various nanostructures, such as nanowires, nanorods, nanoflowers, nanospheres, nanosheets, nanotubes and so on.

Purushothaman and his group [33] successfully prepared NiO nanoparticles via the hydrothermal method, using SDS as a surfactant. The different temperatures (120, 140, 160 and 180°C) in hydrothermal processes have been studied to optimize the morphology and electrochemical properties of NiO. The high degree of phase purity of the NiO particles with nanosizes of 8–16 nm have been achieved under all selected temperatures. However, the morphologies and electrochemical properties were different. At 120 and 140°C, the assembly of nanosheets is slow and, hence, they probably assemble into microspheres under the assistance of a surfactant. A decrease in the surface tension with increasing preparation temperature results in weak electrostatic interaction. The reduced surface tension lowers the aggregation, enabling the formation of microspheres with well-resolved nanosheets at 160°C. The initial nucleation and growth rate will be faster at 180°C. The faster nucleation hinders the assembly of anion surfactant and cation, resulting in the formation of nanorod assembled thicker plates. In case of supercapacitance, the sample prepared at 120°C exhibited a specific capacitance of 871 F g^{-1}, while the value of sample formed at 140°C was 925 F g^{-1}. The maximum specific capacitance of 989 F g^{-1} was obtained at 160°C while the specific capacitance showed a reduced value of 496 F g^{-1} at 180°C. The formation of a nanosheet-like structure seem to have facilitated the ion exchange process by reducing the diffusion lengths for the electrolyte, yielding a superior redox process in the sample prepared at 160°C, which exhibited the best specific capacitance. The specific capacitance was lower at elevated temperatures (180°C) might because the crystallites of larger size were formed, and they limited the paths available for ion transport. Moreover, Xia et al. [34] synthesized hollow Co_3O_4 nanowire arrays through a facile hydrothermal method. The Co_3O_4 nanowires have an average diameter of 200 nm, and the hollow centers have a diameter of 25 nm. In addition, a hierarchically porous can be found in nanowires allowing easier electrolyte penetration. Such a novel structure with porous walls and hollow center possesses more sites for ions to enter and allows facile ion diffusion at high current density leading to superior specific capacitances of 599 F g^{-1} at a current density of 2 A g^{-1} and 439 F g^{-1} at 40 A g^{-1}.

3.2. Electrodeposition methods

Electrochemical deposition is a simple, fast, nonpolluting and facile technique, thus becomes one of most commonly used approaches to prepare metal oxide thin films. An electrochemical synthesis is achieved by a series of procedures that electron transfer between two or more electrodes separated by electrolyte making the occurrence of oxidation or reduction in the electrode–electrolyte interface which finally results in thin films deposited on electrode substrates. Electrochemical deposition can be divided into two different methods: anodic deposition and cathodic deposition. For example, Aghazadeh [35] prepared nanostructured Co_3O_4 via a simple cathodic electrodeposition method. The porous Co_3O_4 nanoplates displayed the average pore diameter and the surface area of 4.75 nm and 208.5 m^2 g^{-1}, respectively. A good specific capacitance as high as 393.6 F g^{-1} at the constant current density of 1 A g^{-1} and an excellent capacity retention (96.5% after 500 charge-discharge cycles) was obtained. Deng et al. [36] have reported that the nanoarchitectured CuO electrodes with a 3D hierarchically porous structure were prepared by an anodic electrodeposition method. An exceptionally large specific capacitance of 880 and 800 F g^{-1} was obtained at scan rates of 10 and 200 mV s^{-1}.

According to the different modes of external power supplying, three main electrochemical deposition techniques including potentiostatic, galvanostatic and pulse period methods have been used by researchers. These different deposition routes with different applied current, potential and time have a crucial impact on the surface morphologies and crystal structures of metal oxide thin films. For example, Lee et al. [37] electrodeposited manganese oxide using three different modes: constant potential (CP) at 1 V for 900 s, pulse potential (PP) at 1 and 0 V with 0.5 s/0.5 s on-off time for 10,000 s, and pulse reverse potential (PRP) at 1 V and −1 V with 0.5 s/0.5 s interval time for 10,000 s. The different deposition times are applied to obtain similar mass loading. The results demonstrated that the different electrodeposition methods have a significant influence on the morphologies of manganese oxide. A traditional bulk film composed of 100 nm particles was prepared by CP mode and exhibited relatively low specific capacitance of 184 F g^{-1} at a scan rate of 10 mV s^{-1}. In case of PP mode, nanostructured MnOx with porous flower petals morphology was obtained and showed a higher specific capacitance of 227.7 F g^{-1}. The highest specific capacitance of 448 F g^{-1} was achieved by PRP mode due to the formation of nanorods with the average diameter of 20 nm which can supply higher specific surface area and faster ion transfer.

3.3. Sol–gel method

The sol–gel technique also attracts significant attention for the synthesis of nanostructured metal oxides because it offers controllable purity, composition, homogeneity of the products. Kim et al. [38] have reported that NiO nanostructures with three distinct morphologies were fabricated by a sol–gel method. The nanoflower structure was created in hexamethylene tetramine (HMTA) solution, while the nanoslice (diameters of 300–530 nm) was prepared in ammonium hydroxide (NH_4OH) solution. The smaller nanoparticles with a diameter of around 50 nm were obtained when the reaction process took place in a strong basic LiOH. In addition, their morphology-dependent supercapacitor properties were exploited. Compared to the nanoslice and nanoparticle-shaped NiO, the nanoflower-shaped NiO showed the best supercapacitor properties (480 F g^{-1} at 0.5 A g^{-1}) despite it having the lowest specific surface area. This is because that the flower-shaped nanostructure has the unique three-dimensional (3D) networks which can provide longer diffusion paths and the highest pore volume (0.66 cm^3 g^{-1}) which offers advantages in contact with and transport of the electrolyte.

Yu and coworkers [39] successfully prepared the three-dimensional (3D) network mesoporous nanostructured α-MnO_2 (MN-α-MnO_2) powders using an inexpensive glucose–permanganate sol–gel method at room temperature and under ambient pressure. The MN-α-MnO_2 exhibited high specific surface areas (ca. 220 m^2 g^{-1}) and narrow pore size distributions (5.6 nm) resulting in a good specific capacitance of 264 F g^{-1} after 1000 charge-discharge cycles.

3.4. Microwave-assisted method

Recently, microwave-assisted method has drawn large attention in the synthesis of oxide materials for supercapacitors application. Compared to conventional oil bath or hydrothermal heating, microwave heating can reduce the reaction time often by orders of magnitude, reduce the manufacturing cost and enhance product yield. An inverted temperature gradient takes place during the microwave-assisted process, and a rapid dielectric heating is generated internally within the material due to applied microwave radiation with a commonly used frequency of 0.3–2.45 GHz. Microwave-assisted synthesis has been adopted to prepare metal oxides with highly uniform nanostructures [40–42].

For example, Zhang et al. [43] have successfully synthesized γ-MnO_2 nanoparticles and α-MnO_2 urchin-like nanostructures by the microwave-assisted reflux as short as 5 min under neutral and acidic conditions, respectively. The γ-MnO_2 nanoparticles showed a smaller particle size, a higher specific surface area and a larger pore volume than those of α-MnO_2 urchin-like nanostructures resulting in a higher capacitance of 311 F g^{-1} at a current density of 0.2 A g^{-1}. The specific capacitance retention and coulombic efficiency after 5000 cycles at 1 A g^{-1} were about 93% and almost 100% for γ-MnO_2 nanoparticles, respectively.

Cao and coworkers [44] prepared flower-like NiO hollow nanosphere precursors via an efficient gas/liquid interfacial microwave-assisted process and were then transformed to NiO by simple calcinations. The wall of the sphere is composed of twisted NiO nanosheets that intercalated with each other. Such hollow structure is different from widely reported flower-like nanostructures with solid cores. These flower-like NiO hollow nanospheres have high surface area of 176 m^2 g^{-1}. Electrochemical properties show a high specific capacitance of 585 F g^{-1} at a discharge current of 5 A g^{-1} and excellent cycling stability.

3.5. Template-assisted method

Hard templates are those materials which are either used as scaffolds for the deposition or employed not only as shape defined templates, but also as chemical reagents that react with other chemicals to produce desired nanomaterials. In the development of various metal oxide nanostructures, the hard template method is widely used and coupled with other methods, such as electrochemical deposition, solvothermal/hydrothermal and sol–gel methods. There are quite a lot of hard templates have been used for the synthesis of metal oxide nanostructures, such as porous anodic aluminum oxide (AAO), polycarbonate membranes (PC), carbon spheres, porous carbon, SiO_2 spheres, mesoporous silica and naturally existing diatomite.

The anodic aluminum oxide (AAO) film is also one of the attractive templates since it possesses very regular and highly anisotropic porous structures with pore diameters ranging from below 10 to 200 nm, pore length from 1 to 50 mm and pore densities in the range of 109–1011 cm^{-2} [45]. The pores have been found to be uniform and nearly parallel, which is useful for the synthesis of one-dimensional metal oxide nanostructures, affording short ion diffusion paths and fast kinetics during the electrochemical reactions. Using AAO as the template, Dar et al. [46] synthesized NiO nanotubes via electrochemical deposition and nanorods after 25 min annealing at 450°C. Due to a suitable combination of nanocrystalline grain size and the high surface area akin to the tubular structure, NiO nanotube exhibits an excellent supercapacitive performance with a maximum specific capacitance of 2093 F g^{-1} which approaches the theoretical value of NiO (2584 F g^{-1}). In contrast, the NiO nanorod structure is characterized by lower performance (797 F g^{-1}). Furthermore, both NiO nanotube and nanorod show high stabilities with almost no alteration to performance after 500 cycles at high current densities of 125 and 80 A g^{-1}. It has also been reported by Xu et al. [47] that Co_3O_4 nanotubes were successfully prepared via the AAO template method. The Co_3O_4 nanotubes have an average diameter of 300 nm and thickness of 50 nm which mainly controlled by the pore size of the AAO template. A good specific capacitance of 574 F g^{-1} was also obtained at a current density of 0.1 A g^{-1}. However, the high cost of AAO templates limits their large-scale application for the production of well-organized metal oxide nanostructured electrodes.

In addition to the AAO template, carbon-based materials with different structures were also developed to prepare metal oxides. For instance, Du et al. [26] reported that Co_3O_4 hollow spheres composed of numerous small nanocrystals were prepared via one-pot hydrothermal carbonization and calcination method with carbon spheres as templates. The specific capacitance is 470 F g^{-1} at a current density of 1 A g^{-1}, and no obvious capacitance decrease was observed over 1000 cycles of charge and discharge. Moreover, Yao et al. [48] synthesized nanostructured hierarchical mesoporous ribbon-like NiO via a hard-template method combining the calcination process. The mesoporous carbon was used as a hard template to control the structure growth and pore size distribution. A large surface area (147 m^2 g^{-1}) and high pore volume (0.2 cm^3 g^{-1}) were achieved when the molar ratio of Ni/C was 2/5. Notably, the outstanding pseudocapacitive performance was obtained with a high specific capacitance of 1260 F g^{-1} at a 1 A g^{-1} and only 5% deterioration of the initial capacitance after 5000 cycles.

4. Metal oxide-carbon composite electrodes

Preparing metal oxide-carbon composites is one of the most effective approaches to improve the supercapacitive performance of metal oxide electrodes. In such composite structures, the carbon materials with large specific surface area and high electric conductivity can provide the channels for charge transfer and benefit to the rate capability. Among a series of carbon materials, carbon nanotubes (CNTs), carbon nanofibers (CNFs), graphene and carbon nanofoams have been mostly studied to combine with metal oxide.

4.1. Metal oxide-carbon nanotubes (CNTs) composite

CNTs have outstanding pore structure, high electrical conductivity, and good mechanical and thermal stability which make them one of the most widely used carbon materials for supporting metal oxide. Thus, CNTs have been coupled with various metal oxides such as NiO, Co_3O_4, V_2O_5, MnO_2, SnO_2 and CuO to form the metal oxide–CNTs composite electrodes. It has been reported in 2005 that Lee et al. [49] synthesized NiO/CNTs nanocomposite via a hydrothermal method and explored the influences of CNT network existing in NiO. Compared to bare NiO, NiO/CNTs nanocomposite electrode exhibited a more rectangular shape in the CV curve and a smaller IR loss indicating a better supercapacitive performance. The specific capacitance increased from 122 to 160 F g^{-1} at a scan rate of 2 mV s^{-1} with the presence of 10% CNTs. The optimized properties owe to that CNTs can effectively improve the electrical conductivity of NiO and supply more active sites for redox reaction of NiO by increasing its specific surface area. After that, Lin et al. [50] prepared mesoporous sphere NiO nanostructures dispersing on the surface of CNTs and the maximum specific capacitance of 1329 F g^{-1} was observed at a very high current density of 84 A g^{-1}. Gund et al. [51] fabricated highly flexible electrode with NiO/MWCNTs nanohybrid thin films on stainless steel substrate with an excellent specific capacitance of 1727 F g^{-1} at a current density of 5 mA cm^{-2} and good stability (91% retention after 2000 cycles).

The advantages of the metal oxide-CNTs composite were further demonstrated. Cheng et al. synthesized nanocomposites of V_2O_5 nanowires and interpenetrating CNTs via a hydrothermal process. When the nanocomposite contained 33 wt% of the CNTs, the V_2O_5-CNTs showed the best specific capacitance of 530 F g^{-1} which was significantly higher than the V_2O_5 nanowires (146 F g^{-1}). The improved conductivity and the increased specific surface area (from 83 to 125 m^2 g^{-1}) were considered to be responsible for the better properties. Moreover, Wang et al. [52] designed a Co_3O_4@MWCNT nanocable using multiwall carbon nanotubes (MWCNTs) as the core cable. Compared to the pristine Co_3O_4 which has a low specific capacitance less than 130 F g^{-1}, the prepared Co_3O_4@MWCNT nanocable exhibits a better performance with a specific capacitance of 590 F g^{-1} at 15 A g^{-1} and 510 F g^{-1} even at 100 A g^{-1}. Furthermore, many efforts have been made on the preparation of MnO_2-CNTs nanocomposites in order to improve the supercapacitive performance of MnO_2. For example, MnO_2-CNTs composites were prepared through a modified one-pot reaction process by Li and coworkers [53]. This cross-linked MnO_2 nanoflakes-CNTs structure showed a good specific capacitance of 201 F g^{-1} and remarkable cycle stability (no obvious decay after

10,000 cycles). It has also been reported by Chen et al. [54] that MnO_2 nanoparticles were introduced into the inner wall of CNT channels by a wet-chemistry method. The result of electrochemical tests shows that the composite has a much higher specific capacitance of 225 F g^{-1} than MnO_2 with a value of 13 F g^{-1}.

4.2. Metal oxide-carbon nanofibers (CNFs) composite

CNFs are very attractive as the support for metal oxides in the composite electrodes due to their conductive networks with appropriate pore channels. Typically the metal oxides are coated on the CNFs surface to form a core–shell structure in which the CNFs can serves as the physical backbone support and offer the channel for efficient electron and ion transportation. Zhi and coworkers [55] reported the synthesis of CNFs/MnO_2 nanocomposite with a coaxial cable structure as shown in **Figure 5a**. The CNFs with a diameter of 200 nm coated by 4-nm-thick MnO_2 nanowhiskers sheath giving a high specific surface area could be seen in **Figure 5b**. The nanocomposite electrode showed a good specific capacitance of 311 F g^{-1} for the whole electrode and 900 F g^{-1} for the MnO_2 shell at a scan rate of 2 mV s^{-1}. In addition, **Figure 5c** and **d** indicated that this CNFs/MnO_2 nanocomposite also exhibited good cycling stability (2.4% loss after 1000 cycles), high energy density (80.2 Wh kg^{-1}) and high power density (57.7 kW kg^{-1}). Moreover, a Fe_3O_4/CNFs nanocomposite was designed by Fu et al. [56] through a solvent thermal reaction. Compared to the low specific capacitance (4 F g^{-1}) of pure Fe_3O_4, the calculated specific capacitance of Fe_3O_4/CNFs nanocomposite is as high as 127 F g^{-1} which indicates a better supercapacitive performance. The CNFs have not only improved the electronic/ionic conductivity of Fe_3O_4, but also prevented the aggregation of Fe_3O_4 nanosheets. CuO has also been used to fabricate composite with CNFs in order to improve its supercapacitive performance. As reported by Moosavifard [57], one-dimensional hierarchical hybrid CuO nanorod arrays-CNFs composite has been prepared via a solution method and an annealing treatment. The CuO nanorods with a length of around 300 nm and a diameter of around 15 nm are grown uniformly surrounding the CNFs. It should be noted that empty space existed among adjacent nanorods indicating a hierarchical array structure. The unique nanocomposite structure contributed to a high capacitance of 398 F g^{-1}.

In addition, carbon fibers can also form a paper. Ghosh et al. [58] prepared carbon nanofiber paper (CFP) with fiber diameters ranging from 100 to 300 nm by electrospinning the polyacrylonitrile (PAN) precursor. The CFP showed good conductivity (0.1 S cm^{-1}), high porosity and large surface area of 700 m^2 g^{-1} indicating its potential to be a promising material for supporting metal oxide. Using this carbon nanofiber paper as substrate, 3-nm-thick V_2O_5 was obtained by electrodeposition method. The V_2O_5-CFP composite exhibited a total specific capacitance of 214 F g^{-1}. Recently, Yang et al. [59] have reported a CFP/Co_3O_4 paper electrode with an excellent specific capacitance of 1124 F g^{-1} at a high current density of 25.34 A g^{-1} in the NaOH electrolyte. The composite also displayed a remarkable electrochemical stability with around 94.4% retention after 5000 charge-discharge cycles. The outstanding supercapacitive performance was attributed to the unique 1D nanonet structure of the electrodes and the improved electronic conductivity as well as ion diffusion by CFP.

http://dx.doi.org/10.5772/65155

4.3. Metal oxide-graphene composite

Since a mechanically exfoliated graphene monolayer was first observed and characterized in 2004, considerable research has been carried out in supercapacitor applications due to its large theoretical specific area (2630 $m^2\ g^{-1}$), high electrical conductivity (104 S cm^{-2}), abundant raw material resource and good electrochemical stability [60]. Therefore, graphene with these fascinating properties is expected as the potential supporting materials to improve the performance of metal oxide-based supercapacitors. Besides graphene, graphene oxide (GO) and reduced graphene oxide (rGO) have also attracted considerable attention due their unique physical and chemical properties. Until now, various metal oxides such as NiO, MnO_2, CuO, V_2O_5, Co_3O_4 and TiO_2 have been coupled with graphene materials to form supercapacitor electrodes for superior performance.

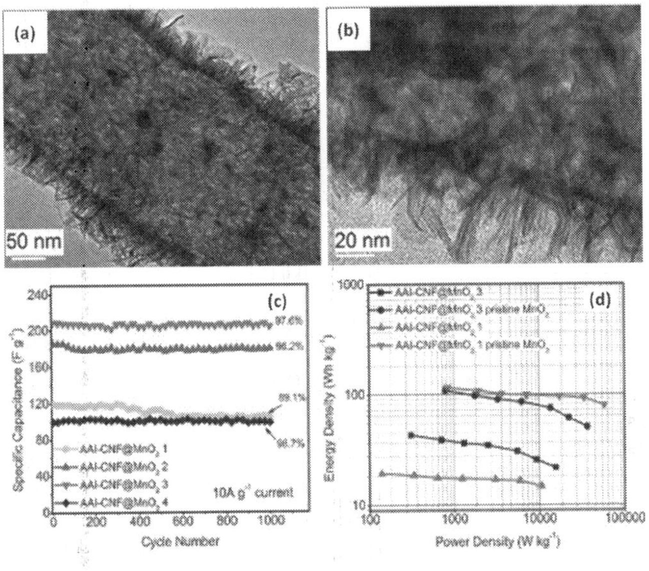

Figure 5. (a) TEM images of a single CNF@MnO_2 nanostructure and (b) the MnO_2 porous shell with nanowhiskers; (c) stability of the CNF@MnO_2 electrodes; (d) Ragone plots of the CNF@MnO_2 electrodes [55].

For example, Ge et al. [61] reported the preparation of 3D flower-like NiO and graphene sheets composite via incorporating a facile hydrothermal process with a thermal treatment process. The resultant composite exhibits a specific capacitance of 346 F g^{-1} (1.5 A g^{-1}), a good rate performance and cycle stability in 2 M KOH. It should be noted that NiO in the composite could provide a specific capacitance as high as 778.7 F g^{-1}, which far exceeded the bare NiO of only 220 F g^{-1}. The magnitude of equivalent series resistances (ESR) are 0.71, 0.99 and 0.85 Ω cm^{-2} for graphene sheet, NiO and composite, respectively, indicating that the conductivity of the composite is improved by the presence of graphene sheets which could contribute to the superior supercapacitive performance of NiO. Wu et al. [62] successfully synthesized NiO particles/graphene oxide (GO) nanosheets composites. Compared with pure NiO and graphene, the composite electrode showed highest current response during CV scanning, reflecting the high capacitance value. The specific capacitance of the NiO/GO composite electrode (460 F g^{-1}) is much higher than those of the bare graphene oxide electrode (13 F g^{-1}) and NiO (40 F g^{-1}) electrode at a current density of 10 A g^{-1}. In addition, the capacitance retention of NiO/GO composite can remain nearly 100% after 3000 cycles which indicates excellent cycle-life stability. Moreover, Co_3O_4/graphene nanosheet (GNS) composite has been synthesized via a microwave-assisted method by Yan et al. [63]. The Co_3O_4 nanoparticles with a small size of 3–5 nm were uniformly distributed on the surface of graphene sheets. The electrochemical properties of composite were observed with a good specific capacitance of 243.2 F g^{-1} and an outstanding stability (only 4.3% loss after 2000 cycles).

Recently, the uniform rod-like V_2O_5 nanocrystals have been fabricated on the surface of reduced graphene oxide (rGO) to form the V_2O_5-rGO nanocomposites as the supercapacitor electrode [64]. The V_2O_5 nanoparticles on rGO were prepared by hydrolysing vanadium oxytripropoxide (VOTP) in ethanol solution with the existence of GO. For a comparative study, the pure V_2O_5 was also prepared in the same condition but without GO. The results have shown that the electrochemical performance of the V_2O_5-rGO nanocomposites with an excellent specific capacitance of 537 F g^{-1} at a current density of 1 A g^{-1} was much better than pure V_2O_5 which had a relatively low value of 202 F g^{-1}. After 1000 charge-discharge cycles, the composite electrode could retain 84% of its initial capacitance while only 30% was retained for pure V_2O_5 indicating that V_2O_5-rGO nanocomposite electrode had a better electrochemical stability. In addition, the higher power and energy densities were also obtained in the composite. The synergistic effect of V_2O_5 nanorods and rGO has been considered to be responsible for the better supercapacitive performance. Firstly, the conductivity can be improved due to the presence of rGO. Secondly, the nanocomposites possess a larger surface area (49.16 m^2 g^{-1}) than that of pure V_2O_5 (37.57 m^2 g^{-1}). Thirdly, the rGO sheets can inhibit the disintegration of V_2O_5 and buffer the strain aroused by the volume expansion during the charging and discharging processes. Finally, the strong adhesion between V_2O_5 nanorods and rGO sheets may facilitate fast electron transfer through the highly conductive rGO sheets. Similarly, Xiang et al. [65] fabricated the rGO–TiO_2 nanobelt composite by a hydrothermal processing in the ethanol solution. When the rGO/TiO_2 mass ratio was 7:3, the composite obtained the best specific capacitance of 200 F g^{-1} which far exceeded pure TiO_2 nanobelt (17 F g^{-1}) and rGO (40 F g^{-1}) in the Na_2SO_4 electrolyte at a scan rate of 2 mV s^{-1}.

4.4. Metal oxide-carbon foams composite

In addition to high specific capacitance, high energy density and power density are also desirable for metal oxide-based supercapacitors. In general, increasing the mass loading can effectively store more energy and power. However, it is a challenge to load a large amount of materials on electrode without undermining the electrochemical performance. One promising approach is to form metal oxide-carbon nanofoams composite as electrodes. Three-dimensional (3D) carbon nanofoams with a through-connected pore network possess large specific surface areas allowing high metal oxide loading. Moreover, the high electric conductivity of carbon nanofoams can improve the electrochemical properties of metal oxide. Chen et al. [66] fabricated nanostructured MnO_2 on CNT foams (or sponges) and the flower-like MnO_2 nanoparticles were uniformly deposited on the skeleton of CNT sponges. An outstanding specific capacitance of 1270 F g^{-1} close to the theoretical value has been obtained and only 4% of degradation after 10,000 cycles at a charge-discharge current density of 5 A g^{-1}. Furthermore, the specific power and energy of this composite are high with values of 63 kW kg^{-1} and 31 Wh kg^{-1}, respectively. Dong and co-workers deposited Co_3O_4 nanowires on the 3D graphene foam in [67]. It can be seen in the figure, the graphene skeleton is fully and uniformly covered by the network of Co_3O_4 nanowires together provide a large accessible surface area. The composite electrode exhibited a high specific capacitance of ~1100 F g^{-1} after 500 cycles at a current density of 10 A g^{-1} and stayed stable afterward indicating a good cycling stability.

5. Conclusion

This chapter presents a relatively general understanding of the correlation between the composition, microstructure and electrochemical behaviors of metal oxide nanostructures-based electrodes for the applications of supercapacitors. The current possibility of controlled growth and self-assembly represents an important step toward the design and tuning of metal oxide nanocrystals, and also it will be a significant step to the applications of metal oxides as ideal electrodes in high performance electrochemical energy storage devices.

Author details

Zhenjun Qi*, Shihao Huang, Adnan Younis, Dewei Chu and Sean Li

*Address all correspondence to: zhenjun.qi@student.unsw.edu.au

School of Materials Science and Engineering, University of New South Wales, Sydney, NSW, Australia

References

[1] Conway, B.E., Transition from "supercapacitor" to "battery" behavior in electrochemical energy storage. Journal of the Electrochemical Society, 1991. 138(6): p. 1539–1548.

[2] Kandalkar, S., et al., Chemical synthesis of cobalt oxide thin film electrode for supercapacitor application. Synthetic Metals, 2010. 160(11): p. 1299–1302.

[3] Zhang, L.L. and X. Zhao, Carbon-based materials as supercapacitor electrodes. Chemical Society Reviews, 2009. 38(9): p. 2520–2531.

[4] Simon, P. and Y. Gogotsi, Materials for electrochemical capacitors. Nature Materials, 2008. 7(11): p. 845–854.

[5] Jayalakshmi, M. and K. Balasubramanian, Simple capacitors to supercapacitors-an overview. International Journal of Electrochemical Science, 2008. 3(11): p. 1196–1217.

[6] Padhi, A.K., K. Nanjundaswamy, and J. Goodenough, Phospho-olivines as positive-electrode materials for rechargeable lithium batteries. Journal of the Electrochemical Society, 1997. 144(4): p. 1188–1194.

[7] Kötz, R. and M. Carlen, Principles and applications of electrochemical capacitors. Electrochimica Acta, 2000. 45(15): p. 2483–2498.

[8] Zhang, S. and G.Z. Chen, Manganese oxide based materials for supercapacitors. Energy Materials, 2008. 3(3): p. 186–200.

[9] Sharma, P. and T. Bhatti, A review on electrochemical double-layer capacitors. Energy Conversion and Management, 2010. 51(12): p. 2901–2912.

[10] Babakhani, B. and D.G. Ivey, Improved capacitive behavior of electrochemically synthesized Mn oxide/PEDOT electrodes utilized as electrochemical capacitors. Electrochimica Acta, 2010. 55(12): p. 4014–4024.

[11] Huang, M., et al., MnO_2-based nanostructures for high-performance supercapacitors. Journal of Materials Chemistry A, 2015. 3(43): p. 21380–21423.

[12] Burke, A., Ultracapacitors: why, how, and where is the technology. Journal of Power Sources, 2000. 91(1): p. 37–50.

[13] Vangari, M., T. Pryor, and L. Jiang, Supercapacitors: review of materials and fabrication methods. Journal of Energy Engineering, 2012. 139(2): p. 72–79.

[14] Conway, B.E., Electrochemical capacitors based on pseudocapacitance. In: Electrochemical Supercapacitors. Springer US; 1999. p. 221–257.

[15] Zhi, M., et al., Nanostructured carbon–metal oxide composite electrodes for supercapacitors: a review. Nanoscale, 2013. 5(1): p. 72–88.

[16] Zheng, J., P. Cygan, and T. Jow, Hydrous ruthenium oxide as an electrode material for electrochemical capacitors. Journal of the Electrochemical Society, 1995. 142(8): p. 2699–2703.

[17] Wei, J., N. Nagarajan, and I. Zhitomirsky, Manganese oxide films for electrochemical supercapacitors. Journal of Materials Processing Technology, 2007. 186(1): p. 356–361.

[18] Devaraj, S. and N. Munichandraiah, Effect of crystallographic structure of MnO_2 on its electrochemical capacitance properties. The Journal of Physical Chemistry C, 2008. 112(11): p. 4406–4417.

[19] Brock, S.L., et al., A review of porous manganese oxide materials. Chemistry of Materials, 1998. 10(10): p. 2619–2628.

[20] Brousse, T., et al., Crystalline MnO_2 as possible alternatives to amorphous compounds in electrochemical supercapacitors. Journal of the Electrochemical Society, 2006. 153(12): p. A2171-A2180.

[21] Wang, G., L. Zhang, and J. Zhang, A review of electrode materials for electrochemical supercapacitors. Chemical Society Reviews, 2012. 41(2): p. 797–828.

[22] Gao, Y., et al., Electrochemical capacitance of Co_3O_4 nanowire arrays supported on nickel foam. Journal of Power Sources, 2010. 195(6): p. 1757–1760.

[23] Lu, Z., et al., Stable ultrahigh specific capacitance of NiO nanorod arrays. Nano Research, 2011. 4(7): p. 658–665.

[24] Huang, M., et al., Self-assembly of mesoporous nanotubes assembled from interwoven ultrathin birnessite-type MnO_2 nanosheets for asymmetric supercapacitors. Scientific Reports, 2014. 4: p. 3878.

[25] Yan, X., et al., Rational synthesis of hierarchically porous NiO hollow spheres and their supercapacitor application. Materials Letters, 2013. 95: p. 1–4.

[26] Du, H., et al., Facile carbonaceous microsphere templated synthesis of Co_3O_4 hollow spheres and their electrochemical performance in supercapacitors. Nano Research, 2013. 6(2): p. 87–98.

[27] Moosavifard, S.E., et al., Designing 3D highly ordered nanoporous CuO electrodes for high-performance asymmetric supercapacitors. ACS Applied Materials & Interfaces, 2015. 7(8): p. 4851–4860.

[28] Wang, Y., et al., Synthesis of 3D-nanonet hollow structured Co_3O_4 for high capacity supercapacitor. ACS Applied Materials & Interfaces, 2014. 6(9): p. 6739–6747.

[29] Qi, Z., et al., A facile and template-free one-pot synthesis of Mn_3O_4 nanostructures as electrochemical supercapacitors. Nano-Micro Letters, 2016. 8(2): p. 165–173.

[30] Nandy, S., et al., Enhanced p-type conductivity and band gap narrowing in heavily Al doped NiO thin films deposited by RF magnetron sputtering. Journal of Physics: Condensed Matter, 2009. 21(11): p. 115804.

[31] Hamdani, M., R. Singh, and P. Chartier, Co_3O_4 and Co-based spinel oxides bifunctional oxygen electrodes. International Journal of Electrochemical Science, 2010. 5(4): p. 556.

[32] Yang, J., et al., Nanostructured porous MnO_2 on Ni foam substrate with a high mass loading via a CV electrodeposition route for supercapacitor application. Electrochimica Acta, 2014. 136: p. 189–194.

[33] Purushothaman, K.K., et al., Nanosheet-assembled NiO microstructures for high-performance supercapacitors. ACS Applied Materials & Interfaces, 2013. 5(21): p. 10767–10773.

[34] Xia, X.-h., et al., Self-supported hydrothermal synthesized hollow Co_3O_4 nanowire arrays with high supercapacitor capacitance. Journal of Materials Chemistry, 2011. 21(25): p. 9319–9325.

[35] Aghazadeh, M., et al., A facile route to preparation of Co_3O_4 nanoplates and investigation of their charge storage ability as electrode material for supercapacitors. Journal of Materials Science: Materials in Electronics, 2016. 27(8) p. 1–10.

[36] Deng, M.-J., et al., Facile electrochemical synthesis of 3D nano-architectured CuO electrodes for high-performance supercapacitors. Journal of Materials Chemistry A, 2014. 2(32): p. 12857–12865.

[37] Lee, S.H., et al., Morphology and composition control of manganese oxide by the pulse reverse electrodeposition technique for high performance supercapacitors. Journal of Materials Chemistry A, 2013. 1(46): p. 14606–14611.

[38] Kim, S.-I., et al., Facile route to an efficient NiO supercapacitor with a three-dimensional nanonetwork morphology. ACS Applied Materials & Interfaces, 2013. 5(5): p. 1596–1603.

[39] Yu, L.L., J.J. Zhu, and J.T. Zhao, Three-dimensional network mesoporous nanostructured α-manganese dioxide with high supercapacitive performance: facile, environmental and large-scale synthesis. European Journal of Inorganic Chemistry, 2013. 2013(21): p. 3719–3725.

[40] Bilecka, I. and M. Niederberger, Microwave chemistry for inorganic nanomaterials synthesis. Nanoscale, 2010. 2(8): p. 1358–1374.

[41] Truong, T.T., et al., Morphological and crystalline evolution of nanostructured MnO_2 and its application in lithium–air batteries. ACS Nano, 2012. 6(9): p. 8067–8077.

[42] Pahalagedara, L., et al., Microwave-assisted hydrothermal synthesis of α-MnO_2: lattice expansion via rapid temperature ramping and framework substitution. The Journal of Physical Chemistry C, 2014. 118(35): p. 20363–20373.

[43] Zhang, X., et al., Microwave-assisted reflux rapid synthesis of MnO_2 nanostructures and their application in supercapacitors. Electrochimica Acta, 2013. 87: p. 637–644.

[44] Cao, C.-Y., et al., Microwave-assisted gas/liquid interfacial synthesis of flowerlike NiO hollow nanosphere precursors and their application as supercapacitor electrodes. Journal of Materials Chemistry, 2011. 21(9): p. 3204–3209.

[45] Liu, N.W., et al., Fabrication of anodic-alumina films with custom-designed arrays of nanochannels. Advanced Materials, 2005. 17(2): p. 222–225.

[46] Dar, F.I., K.R. Moonoosawmy, and M. Es-Souni, Morphology and property control of NiO nanostructures for supercapacitor applications. Nanoscale Research Letters, 2013. 8(1): p. 1.

[47] Xu, J., et al., Preparation and electrochemical capacitance of cobalt oxide (Co_3O_4) nanotubes as supercapacitor material. Electrochimica Acta, 2010. 56(2): p. 732–736.

[48] Yao, M., et al., Template synthesis and characterization of nanostructured hierarchical mesoporous ribbon-like NiO as high performance electrode material for supercapacitor. Electrochimica Acta, 2015. 158: p. 96–104.

[49] Lee, J.Y., et al., Nickel oxide/carbon nanotubes nanocomposite for electrochemical capacitance. Synthetic Metals, 2005. 150(2): p. 153–157.

[50] Lin, P., et al., The nickel oxide/CNT composites with high capacitance for supercapacitor. Journal of the Electrochemical Society, 2010. 157(7): p. A818-A823.

[51] Gund, G.S., et al., Architectured morphologies of chemically prepared NiO/MWCNTs nanohybrid thin films for high performance supercapacitors. ACS Applied Materials & Interfaces, 2014. 6(5): p. 3176–3188.

[52] Wang, X., et al., Co_3O_4@ MWCNT nanocable as cathode with superior electrochemical performance for supercapacitors. ACS Applied Materials & Interfaces, 2015. 7(4): p. 2280–2285.

[53] Li, L., et al., Facile synthesis of MnO_2/CNTs composite for supercapacitor electrodes with long cycle stability. The Journal of Physical Chemistry C, 2014. 118(40): p. 22865–22872.

[54] Chen, W., et al., Enhanced capacitance of manganese oxide via confinement inside carbon nanotubes. Chemical Communications, 2010. 46(22): p. 3905–3907.

[55] Zhi, M., et al., Highly conductive electrospun carbon nanofiber/MnO_2 coaxial nanocables for high energy and power density supercapacitors. Journal of Power Sources, 2012. 208: p. 345–353.

[56] Fu, C., A. Mahadevegowda, and P.S. Grant, Fe_3O_4/carbon nanofibres with necklace architecture for enhanced electrochemical energy storage. Journal of Materials Chemistry A, 2015. 3(27): p. 14245–14253.

[57] Moosavifard, S.E., et al., Facile synthesis of hierarchical CuO nanorod arrays on carbon nanofibers for high-performance supercapacitors. Ceramics International, 2014. 40(10): p. 15973–15979.

[58] Ghosh, A., et al., High pseudocapacitance from ultrathin V_2O_5 films electrodeposited on self-standing carbon-nanofiber paper. Advanced Functional Materials, 2011. 21(13): p. 2541–2547.

[59] Yang, L., et al., Hierarchical network architectures of carbon fiber paper supported cobalt oxide nanonet for high-capacity pseudocapacitors. Nano Letters, 2011. 12(1): p. 321–325.

[60] Stoller, M.D., et al., Graphene-based ultracapacitors. Nano Letters, 2008. 8(10): p. 3498–3502.

[61] Ge, C., et al., Three-dimensional flower-like nickel oxide supported on graphene sheets as electrode material for supercapacitors. Journal of Sol-Gel Science and Technology, 2012. 63(1): p. 146–152.

[62] Wu, M.-S., et al., Formation of nano-scaled crevices and spacers in NiO-attached graphene oxide nanosheets for supercapacitors. Journal of Materials Chemistry, 2012. 22(6): p. 2442–2448.

[63] Yan, J., et al., Rapid microwave-assisted synthesis of graphene nanosheet/Co_3O_4 composite for supercapacitors. Electrochimica Acta, 2010. 55(23): p. 6973–6978.

[64] Li, M., et al., Controlling the formation of rodlike V_2O_5 nanocrystals on reduced graphene oxide for high-performance supercapacitors. ACS Applied Materials & Interfaces, 2013. 5(21): p. 11462–11470.

[65] Xiang, C., et al., Reduced graphene oxide/titanium dioxide composites for supercapacitor electrodes: shape and coupling effects. Journal of Materials Chemistry, 2012. 22(36): p. 19161–19167.

[66] Chen, W., et al., High-performance nanostructured supercapacitors on a sponge. Nano Letters, 2011. 11(12): p. 5165–5172.

[67] Dong, X.-C., et al., 3D graphene–cobalt oxide electrode for high-performance supercapacitor and enzymeless glucose detection. ACS Nano, 2012. 6(4): p. 3206–3213.

Chapter 5

$BaTiO_3$:H Films as All-Solid-State Electrolytes for Integrated Electric Double-Layer Capacitors

Fadhel El Kamel

Additional information is available at the end of the chapter

http://dx.doi.org/10.5772/65110

Abstract

In the electric double-layer capacitors (EDLCs), a large amount of electrical energy can be stored in the double layer by reversible accumulation of ions onto the active electrode material. In these devices, mobile charge carriers can accumulate (or deplete) near the electrode/electrolyte interface resulting in a space charge layer. So, the appropriate combination of space charge layer and large effective surface of the electrodes constitutes a significant factor to get high specific capacitance. Here, we incorporated protons in $BaTiO_3$ films during a low-temperature deposition process. Drastic changes occurred on both chemical and electrical properties of the films when H_2 was added to the sputtering gas. It is well known that protons are very mobile species even at low temperature. Therefore, upon the application of a sufficiently high electric field, positively charged protons move toward the cathode with an activation energy around 0.6 eV and pileup to form a capacitive double layer of several μF/cm^2 which enhances the dielectric permittivity of the film.

Keywords: hydrogenated $BaTiO_3$ films, double layer, all-solid-state supercapacitors

1. Introduction

Owing to the progress in thin-film technology and materials engineering, electronic chips now integrate several functions on the same area, especially in wireless sensor networks, portable equipments, and other microsystems. As a result, microenergy sources need to be developed in order to drive such integrated electronic devices or to provide power during the temporary failure of the primary power sources. Two major performance criteria for any electrical energy storage device are required, energy and power densities. The first is defined as the amount of energy stored per unit mass (Wh/kg) or per unit volume (Wh/m^3) in the device. The latter is a measurement of how fast energy is extracted from, or transferred into, the device per unit mass (W/kg) or per unit volume (W/m^3). The two criteria are particularly important where there is an excessive requirement for portability. Therefore, supercapacitors and Li-ion batteries should not necessarily be seen as competitors, because their charge storage mechanisms and thus their characteristics are different. Batteries convert chemical energy directly to electrical energy, in which charge is generated by redox reaction at electrodes and voltage is established between cell terminals depending on the chemical species and their concentrations. In previous batteries, this energy transformation is irreversible, but in novel ones, the chemical reaction is reversible, and thus the battery can be charged by supplying electrical energy to the cell.

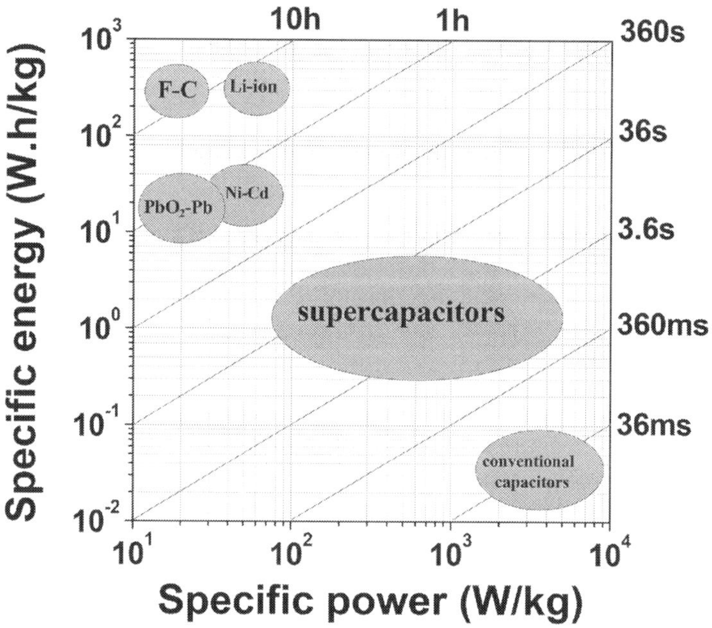

Figure 1. Specific power vs. specific energy (*Ragone plot*) for various electrical energy storage devices. Time constant values of each device are obtained by dividing the energy density by the power density.

Generally, the capacitance obtained with conventional capacitors finds its origin in the electronic, ionic, and dipolar polarization occurring in the bulk, whereas the essential reasons for electrical energy storage in supercapacitors are achieved by both the heavy load of electrode material per unit area and the relatively large specific capacitance of the electrolyte material. Supercapacitors can principally be classified as either electric double-layer capacitors (EDLCs) or pseudocapacitors. The energy storage mechanism in the former devices relies on the separation of charges at the electrode/electrolyte interface, whereas in the latter, a faradic process occurs in addition to a simple charge separation. This feature explains why the charge storage capacity of pseudocapacitors is typically larger than that of EDLCs.

The availability of the stored charge will always be faster for supercapacitor (*surface storage*) than for a Li-ion battery (*bulk storage*), with a larger stored energy for the latter [1]. Both devices must be used in their respective time-constant domains (**Figure 1**).

Using a Li-ion battery for repeated high-power delivery/uptake applications for a short duration (less than 10 s) will quickly degrade the cycle life of the system. The only way to avoid this is to oversize the battery, increasing the cost and volume. In the same way, using supercapacitors for power delivery longer than 10 s requires oversizing.

Supercapacitors [2, 3], until now consisting of liquid-state electrolytes, have been widely regarded as energy storage devices for several electronic systems. Hence, they can ensure this power request since they have high-power density, which can be supplied in a very short time. Pseudocapacitive metal oxides (RuO_2) [4, 5], conducting polymers [6], and carbon-based materials such as carbon nanotubes [7], graphite oxides [8], onion-like carbon [9], and activated carbon [10] have been widely reported in literature as electrode materials especially for liquid-state supercapacitors with interdigital fingers [4, 9, 11] and roll-like [12] and sandwich [8] shapes. To date, these supercapacitors are principally based on liquid-state electrolytes, such as aqueous or organic solutions, and display high capacitance values (from 1 to 100 mF/cm^2), according to the electrode nature and its geometry, the number of interdigitated microelectrodes in the device, the electrolyte, and so on. However, these devices cannot be used at high temperatures, because the aqueous or organic electrolytes undergo decomposition. In addition, they exhibit leakage current and ionic conductivity which vary significantly with temperature and frequency.

Although solid electrolytes are characterized by lower ionic conductivity compared to their liquid counterparts, we can overcome this drawback by decreasing the thickness of solid electrolytes [13] in order to reduce the diffusion path of charged ionic species or by increasing the effective electrode surface by using porous materials [14]. On the other hand, with solid-state electrolyte, supercapacitors have wide operational temperature range and negligible leakage current. Hydrogel-polymer electrolyte has been reported by Kaempgen et al. [15] for the all-solid-state supercapacitors. The estimated cell capacitance was around 1 mF/cm^2, but the operating voltage range is limited to 1 V, making them nonfunctional for most applications. Yoon et al. [5] have investigated the use of a pseudocapacitive amorphous RuO_2 electrode with a $Li_xPO_yN_z$ (LiPON) solid electrolyte for the elaboration of a thin-film supercapacitor. They reported a specific capacitance of about 3.5 $mF/cm^2 \times \mu m$ and a high operating voltage range. However, the fast capacitance degradation observed after several charge/discharge cycles derives from the lower ion mobility of Li^+ ions in the LiPON electrolyte than that of H^+ and OH^- ions in the liquid electrolyte. This suggests that solid electrolyte with high protonic conductivity holds promise for thin-film supercapacitor with high capacitance and high cycle life. Recently, solid-state electrolytes such as hydrated lithium fluoride [2], yttria-stabilized zirconia [3], and Ta_2O_5:H [16] are examples of ionic conductors investigated for such a purpose.

The deposition of solid-state supercapacitors reported up to now involves high-temperature processing [5, 17] or humidified environment [2]. This constitutes a serious limitation to integrate supercapacitors in electronic chips. As an alternative, here protons were incorporated in barium titanate ($BaTiO_3$) films during a low-temperature deposition process. The main purpose of this study was focused on related electrical defects. We emphasize that drastic changes occur on both chemical and electrical properties of the films when H_2 is added to the sputtering gas. The electric double-layer capacitance can reach values up to several $\mu F/cm^2$.

2. Experimental details

Barium titanate films were grown by rf magnetron sputtering process on gold, copper, and carbon nanowalls/Pt-coated silicon substrates (*previously coated with a chromium adhesion layer*). Film thickness is around 1 μm as measured by a Veeco Dektak 6 M surface profiler and confirmed by a scanning electron microscopy. During deposition the substrates were water cooled (*room temperature deposition*). Standard sputtering was performed with pure argon. Here, hydrogen was incorporated in the $BaTiO_3$ layers during their growth by introducing H_2 in the sputtering gas [$H_2/(H_2 + Ar)$ varies from 0 to 30 %]. The stoichiometry of the sputtering gas ($H_2 + Ar$) was controlled using respective mass flows for each gas (Ar *and* H_2). Since the films are grown at low temperature, they are found to be amorphous, as detected by X-ray diffraction.

Electrical and dielectric measurements were performed, in a dark-shielded cell filled with dry nitrogen, on M/*a*-$BaTiO_3$:H/M (M = Au, Cu) planar capacitors, where gold and copper electrodes (1.77 mm^2 *area*) were evaporated through a shadow mask on the front side of the deposited films. Dielectric properties of these devices were studied as a function of temperature and frequency using a Novocontrol impedance analyzer (*with an ac test voltage of* 100 mV).

The capacitance (C), conductance (G), and dissipation factor ($\tan\delta = G/C\omega$) were measured from 10^{-1} to 10^{6} Hz. Known the geometrical factor (*top electrode diameter and film thickness*), the dielectric constant ε' and conductivity σ can be determined, and results will be presented as $\varepsilon'(f)$, $\tan\delta(f)$, and $\sigma(f)$ plots. Current-voltage measurements were performed using a Keithley 6517A electrometer. Temperature variation was controlled by a Linkam hot stage.

3. Results and discussion

Films grown under pure argon are transparent [18]. This transparency diminishes widely with the addition of hydrogen in the sputtering gas. The color of hydrogenated films tends gradually to yellow and then to brown, and finally they darken for hydrogen mixing ratio (HMR) around 25 %. This dark color indicated a high oxygen deficiency [19, 20]. In order to better understand this color change, optical absorbance spectrum was recorded on the hydrogenated barium titanate films at room temperature using a Shimadzu (UV-3101 PC) UV-Vis-NIR scanning spectrophotometer in diffuse reflection mode. Amorphous barium titanate is reported as a direct band gap material. So, it is evident that in the shorter wavelength region, the absorption coefficient α follows a power law:

$$\alpha h\nu = B\sqrt{h\nu - E_0} \qquad (1)$$

where $h\nu$ is the energy of the incident photon, B is the absorption edge width parameter, and E_0 is the optical band gap. Tauc plots $[(\alpha h\nu)^2$ vs. $h\nu]$ for $BaTiO_3$:H films are shown in **Figure 2**.

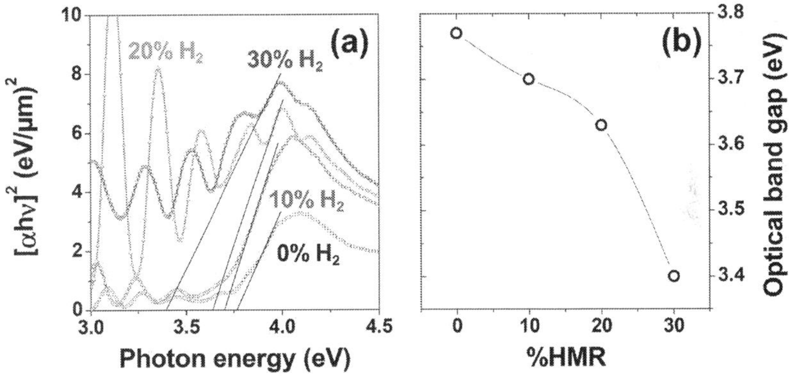

Figure 2. (a) UV-Vis absorbance spectra of hydrogenated barium titanate films deposited under different HMRs. Optical band gap energy was extracted by linear fit calculation. (b) Optical band gap energy as a function of HMR in the sputtering gas.

The linear fit of the straight-line portion of the data indicates a direct band gap with an energy value varying from 3.76 to 3.40 eV when HMR varies from 0 to 30 %. The optical absorption in the UV region is mainly attributed to the electron transition from the valence band maximum to the conduction band minimum. So, a plausible explication for the band gap narrowing can be the existence of several defect levels, oxygen vacancies, or other kinds of punctual defects, within the band gap.

Evidences for hydrogen incorporation were previously given by the Fourier transform infrared spectroscopy and X-ray photoelectron spectroscopy [21]. It is worth noting that the hydrogenated films display a large density of hydroxide compared to the standard films (*grown under pure argon*), while the density of oxide remains constant in both films. The incorporation of hydrogen in the films during the deposition process can explain this feature. As it is already established [22–24], the hydrogen ionizes to give an electron (*donor defect*) and a proton (H^+) which combines with oxygen of the perovskite lattice to form hydroxide groups.

Figure 3 shows an overview on the frequency spectra associated to the capacitance density $C(f)$ (**Figure 3(a)**) and the conductivity $\sigma(f)$ (**Figure 3(b)**) performed at room temperature on a-$BaTiO_3$:H films grown under different hydrogen mixing ratios (HMRs) in the sputtering gas.

We note that as hydrogen was introduced with a content exceeding 10 %, films display (**Figure 3(a)**) hundred times higher permittivity (~2500) than that measured on films grown under pure argon (~20). The double layer is clearly evidenced, especially at low temperatures by a dispersive behavior of the capacitance accompanied by a relaxation peak (LT Relax) in the loss measurements (**Figure 4(a)**).

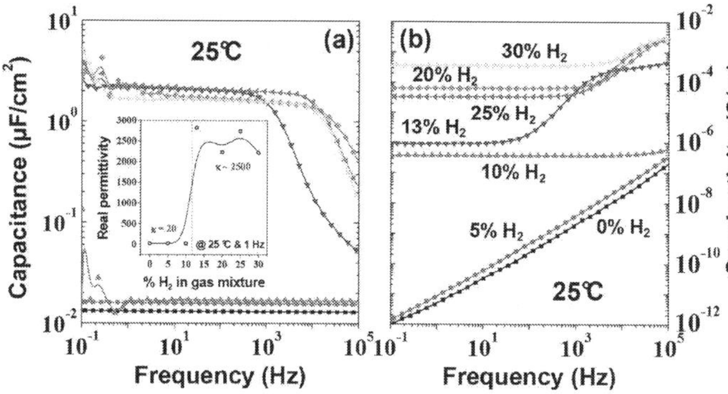

Figure 3. Frequency dependence of the capacitance density (a) and the conductivity (b) measured at room temperature on barium titanate layers sputtered under different hydrogen mixing ratios ranging from 0 to 30 %. The inset shows the variation of real permittivity vs. the hydrogen mixing ratios.

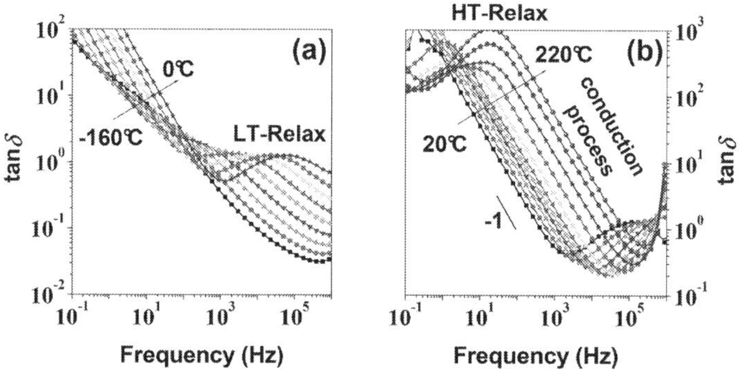

Figure 4. Temperature dependence of the loss factor tan$\delta(f)$, measured on films grown under 25 % HMR. Data curve shows two relaxation peaks, which appear at low (LT Relax). (a)) and high (HT Relax. (b)) temperatures.

We believe that the dielectric losses are mainly the result of the ion migration and/or trap release. At high frequencies (f > 10 kHz), C relaxes toward the bulk contribution, where mobile species cannot reach the electrodes under the ac test voltage (electrode polarization is basically a slow relaxation mechanism). As we can see in **Figure 3(a)**, the bulk capacitance of the hydrogenated films tends to the value measured when films were grown under pure argon.

The cutoff frequency (f = 10 kHz) represents the speed of the capacitor [2] which constitutes an important parameter for both fundamental studies and technological applications. Furthermore, conductivity is also affected by the addition of hydrogen in the sputtering gas. At low frequency, it increases by up to eight orders of magnitude. The $\sigma(f)$ characteristic shifts toward higher frequencies with the appearance of two plateaus, corresponding to the bulk and interfacial conductivities, that appear at high and low frequencies, respectively. This feature was previously [25] explained through the electrode polarization mechanism [26] which arises from proton accumulation at the cathode over a Debye length and gives rise to a large capacitance [25]. In the low-frequency domain, measured capacitance was completely determined by the charge stored in the double layer. As a result, electric double-layer capacitors have been widely regarded as energy storage devices.

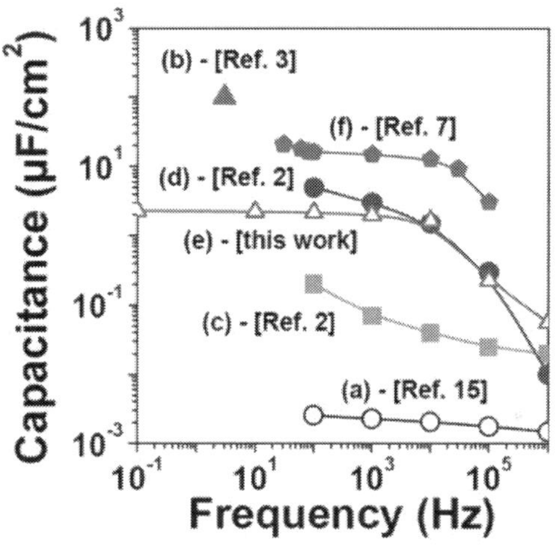

Figure 5. Selected literature data for the $C(f)$ characteristics measured on different capacitors. (a) $BaTiO_3$ crystal (2 mm *thick*) sandwiched between two smooth metal electrodes. Protons are incorporated as mobile carriers by hydrogen charge for 8 days using electrolysis of water. (b) Yttria-stabilized zirconia (0.6 mm *thick*) sandwiched between two porous Pt/YSZ composite layers. Electric double layer is formed by the accumulation of oxygen ions. (c–d) Cu/LiF(0.32 µm *thick*)/Cu-based supercapacitor. Under the electric field, two charge carriers (Li^+ and F^-) are accumulated at the interfaces. Measurements were carried out under two humid environments, 34 % RH (c) and 80 % RH (d). (e) *a*-$BaTiO_3$:H (1 µm *thick*) sandwiched between two smooth Au electrodes (*the present work*). Electric double layer is formed by the accumulation of mobile proton carriers. (f) Solid-state supercapacitor from single-walled carbon nanotube arrays coated with Al_2O_3.

In general, electric double-layer capacitors (EDLCs) exhibit the property that a large electrical energy can be electrostatically stored in the double layer by reversible accumulation of ions (*from the electrolyte*) onto the active electrode material that is electrochemically stable. In these devices mobile charge carriers can accumulate or deplete near the electrode/electrolyte interface resulting in a space charge layer. So, the appropriate combination of space charge layer and large effective surface of the electrodes constitutes a significant factor to get high specific capacitance. **Figure 5** shows selected literature data for the *C(f)* characteristics measured on different capacitors.

The most common devices at present use carbon-based active materials (*or other porous composite materials*) with high surface area as electrodes in EDLCs. In the present study, we used smooth metallic electrodes (*low surface area*) in order to study the effect of electrolyte without any contribution of the electrode nature in addition to carbon nanowall-coated platinum (CNW/Pt) as a high specific surface electrode. **Figure 6** displays the *C(f)*, *ε(f)*, and *σ(f)* characteristics of Au/*a*-$BaTiO_3$/Au (a), Au/*a*-$BaTiO_3$:H/Au (c), and Au/*a*-$BaTiO_3$:H/Cu (d) devices.

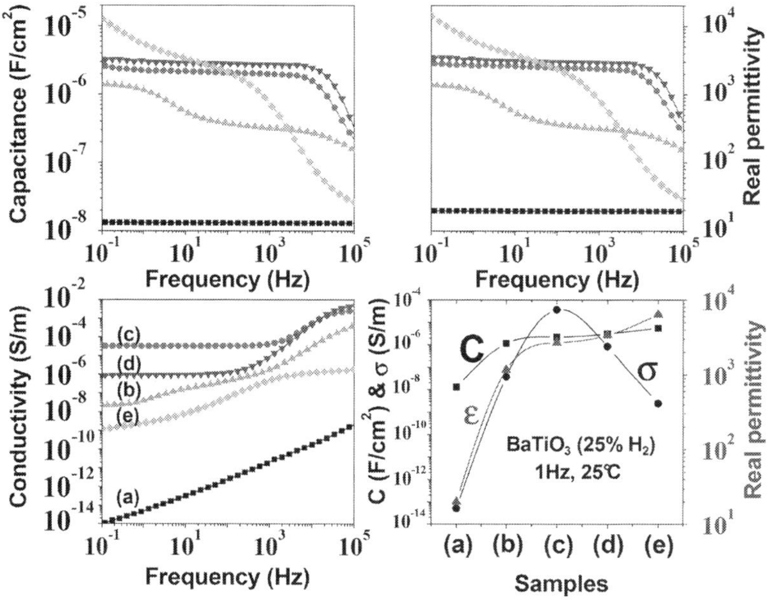

Figure 6. *C(f)*, *ε(f)*, and *σ(f)* characteristics of (a) Au/*a*-$BaTiO_3$/Au, (b) Au/*a*-$BaTiO_3$/*a*-$BaTiO_3$:H(20%H_2)/*a*-$BaTiO_3$/Au, (c) Au/*a*-$BaTiO_3$:H(25%H_2)/Au, (d) Au/*a*-$BaTiO_3$:H(25%H_2)/Cu, and (e) Au/*a*-$BaTiO_3$:H(25%H_2)/CNW/Pt devices. Even with low effective surface of electrodes, we can get acceptable values of capacitance that is higher than 2 μF/cm², which leads to a real permittivity in the 10^3–10^4 range.

We showed that even with low effective surface of electrodes, we can get acceptable values of capacitance that is higher than 2 μF/cm², which leads to a real permittivity in the 10^3–10^4 range. By increasing the surface area of the electrodes, we anticipate higher specific capacitance for our devices as observed with the Au/*a*-$BaTiO_3$:H/CNW/Pt (e) device (C = 10 μF/cm² *and* ε' = 10^4 *at low frequency*).

To study the temperature dependence of the conduction mechanisms taking place in hydrogenated films, we plotted the relaxation frequencies $f_0(T)$, bulk $\sigma_b(T)$, and interfacial $\sigma_i(T)$

conductivities, extracted from the relaxation peak position (**Figure 4**) and high- and low-frequency semicircles in the impedance diagram (**Figure 7**), respectively, as a function of 1000/*T* (*Arrhenius plot*, **Figure 8**).

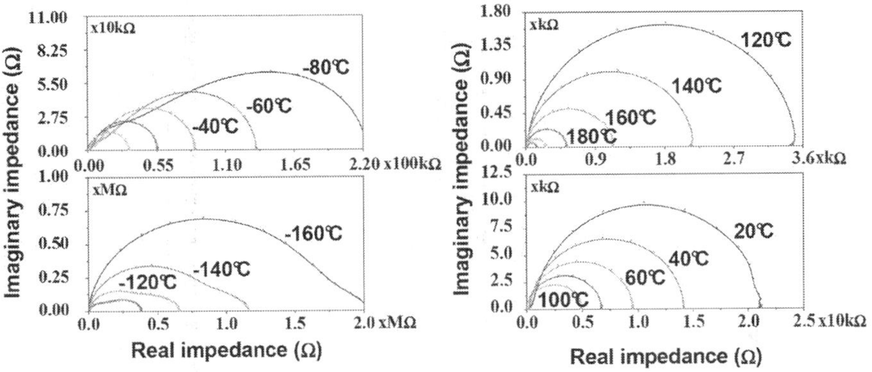

Figure 7. Complex impedance spectra (Z''–Z') carried out, at different temperatures, on hydrogenated barium titanate films. Samples were grown under 25 % HMR in the sputtering gas.

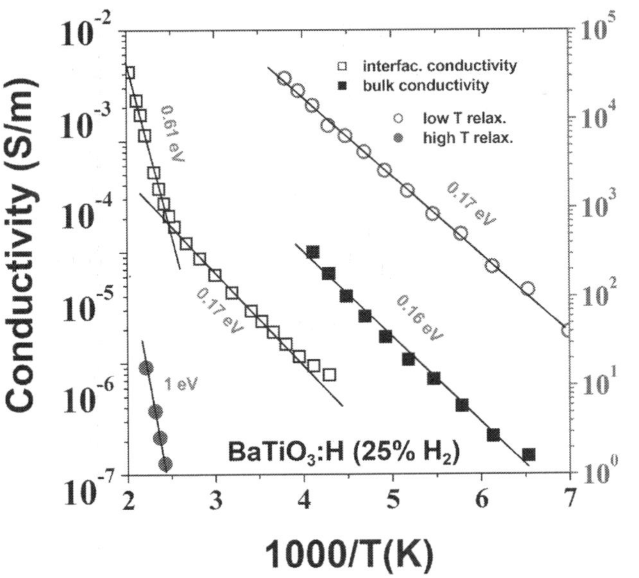

Figure 8. Relaxation frequency, bulk, and interfacial conductivities as a function of 1000/*T* (*Arrhenius diagram*).

Measurements were carried out on samples grown under 25 % HMR in the sputtering gas. Both characteristics show the Arrhenius-type dependence over the whole temperature range, excluding the low-temperature ($T < -100$ °C) interfacial conductivity. As shown in **Figure 8**, we can extract three activation energies which correspond to three different conduction processes.

I–V characteristics were recorded on hydrogenated films at different temperatures (from −150 to −50 °C) in order to explain the conduction mechanism in dc regime. DC bias was applied to the bottom electrode, and the current was measured after 60 s for stabilization concern. Experimental data measured on the Au/*a*-BaTiO3:H (17% HMR)/Au device are shown in **Figure 9(a)**.

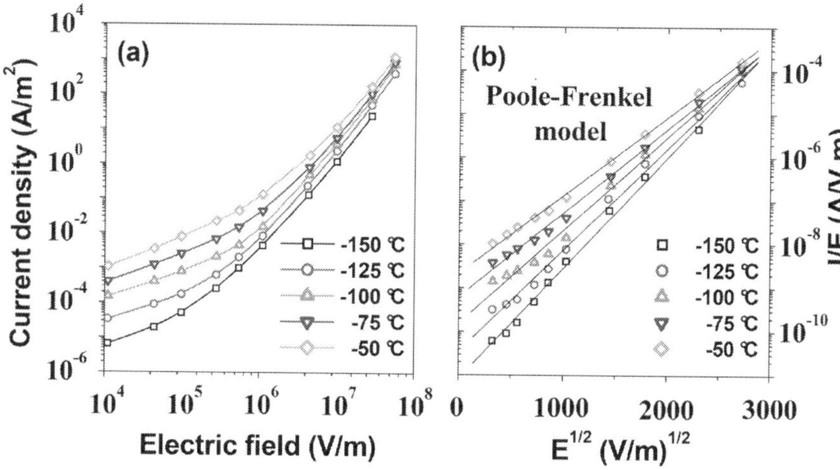

Figure 9. (a) Temperature dependence of the leakage current (*J–E characteristics*). (b) *J–E* characteristics replotted according to the Poole-Frenkel model.

It is clearly seen that the leakage current density exhibits a Poole-Frenkel (PF)-type behavior, which implies that the conduction mechanism can be described by a thermally stimulated emission from a discrete set of traps. The PF model predicts that the current density can be expressed as:

$$J_{\mathrm{PF}} = qN_{\mathrm{c}}\mu E \exp\left(\frac{-q\left(\phi_{\mathrm{t}} - \sqrt{qE/\pi\varepsilon_{\mathrm{opt}}}\right)}{\mathrm{k_B}T} \right) \tag{2}$$

where E is the applied field, $\mathrm{k_B}$ is the Boltzmann constant, μ is the electronic mobility, $\varepsilon_{\mathrm{opt}}$ is the optical permittivity, q is the elementary charge, $q\phi_{\mathrm{t}}$ is the trap level, and N_{c} is the effective density of states in the conduction band assumed to be around 10^{18} cm^{-3} [27]. As stated above

(Poole-Frenkel *model*), log(J/E) vs. $E^{1/2}$ should be linear (**Figure 9(b)**). Then, we guess that the leakage current arises from the carriers release from shallow trap levels localized within the band gap. In order to determine this energy level, we plotted the current density vs. $1/T$ at a fixed electric field. J vs. $1/T$ curves (*not shown here*) display a linear behavior with a negative slope which suggests that the conduction process is thermally stimulated. Activation energy and carrier mobility vary, respectively, from 0.12 and 4.30×10^{-7} to 0.05 eV and 2.00×10^{-6} cm²/Vs when the electric field varies from 0.2 to 3.0 MV/m. Such energy values correspond to an electric-field–dependent effective trap depth [$\varphi_0 - \Delta\varphi(E)$]. Based on the slowness of the PF mechanism, these mobility values are in a good agreement with the electronic conduction. In the literature, electronic mobility shows a large discrepancy and is found to vary from 10^{-3} [28–32] to 10^{-10} cm²/Vs [27]. To determine the real trap's depth φ_0 (*at* $E = 0$ V/m), we plotted the effective trap depths [$\varphi_0 - \Delta\varphi(E)$] as a function of $E^{1/2}$ (**Figure 10**).

We found that the leakage current can be explained by the carrier release from shallow trap level φ_0 localized at around 0.15 eV below the conduction band. It is believed that positively charged protons can provide such energetic levels. It is interesting to note that the same amount of energy is required to activate the LT relaxation, bulk, and interfacial (*at* $T < 120$ °C) conduction processes (**Figure 8**).

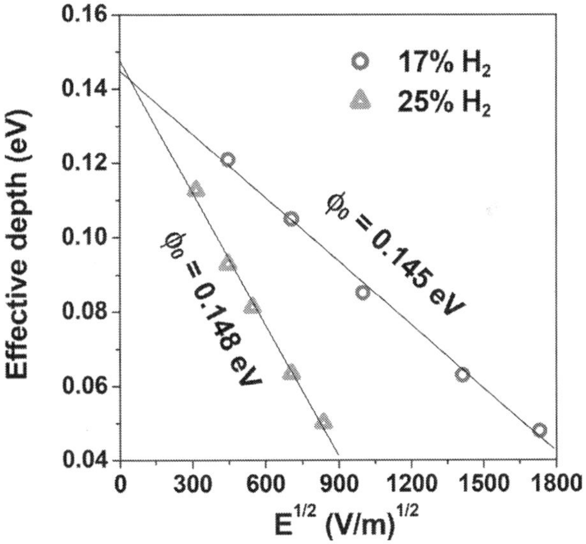

Figure 10. Effective trap depth as a function of $E^{1/2}$. Leakage currents can be explained by carrier detrapping from shallow trap level φ_0 localized at around 0.15 eV below the conduction band.

In amorphous materials, charged defects surrounding the proton tilt markedly the hydroxide group toward one of the neighboring oxygen ions. Oxygen vacancies constitute the main charged defects since the hydrogenated films were deposited under a reduced atmosphere.

They effectively act as positively charged defects that repel the proton. This feature allows the formation of a rather weak directional interaction (*hydrogen bond*) between the proton and one of the adjacent oxygen, $O^{[1]}$-H~$O^{[2]}$ [33]. Such a site could be considered as a trap for protons. In this approach, electron issued from the ionization of hydrogen is weakly bonded, and it can be easily activated to the conduction band, from a shallow donor level estimated at around 0.15 eV, to increase the dielectric loss and the leakage current [34].

The hydrogen bond breakage can be responsible for a large proton conduction process in oxide materials. Several reports (*Ref.* [35] *and references therein*) adopted a model for O-H oscillators localized on regular oxygen sites and undergoing a stretching mode alongside the "O-O" direction. This situation allows small proton oscillations (*displacements to less than atomic spacing*). Due to their weakness, hydrogen bonds require an activation energy around 0.22 eV [36] to be dissociated even at low temperature. This value agrees well with the one (0.26 eV) determined previously [37] using dielectric measurements at temperature ranging from −25 to −75 °C (*not shown here*). Then, the dielectric response rising at that range can be ascribed to the dissociation of hydrogen bonds and consequently to a localized migration (*or oscillation*) of the protons alongside the hydroxide bond. Such process was previously emphasized and discussed by Weber et al. [35] in ionic conductor oxides.

Additionally, the interfacial conduction process considered at temperature higher than 120 °C for hydrogenated films was thermally activated with around 0.6 eV. It is whispered that the proton diffusion within oxide materials requires almost the same amount of energy. Actually, the reported activation energy should be strongly affected by the amorphous state of the material, but it agrees well with predicted experimental [38, 39] and theoretical [38–40] values (0.4–0.6 eV) for the best proton conductors such as Y-$BaZrO_3$ or $BaCeO_3$ [38, 39, 41]. Based on the experimental observation and on the light of literature mentioned above, the conduction mechanism is predominately of the Grotthuss type [42], that is, involving proton transfer (*hopping*) from the hydroxide defects to a nearest neighboring oxygen ion.

In order to better illustrate that the huge increase of the dielectric constant (**Figure 3**) arises from the accumulation of protons at the metal-$BaTiO_3$:H interface (*double layer*), a tri-layer stack ($BaTiO_3$/$BaTiO_3$:H/$BaTiO_3$) was grown by inserting the hydrogen-doped layer between two intrinsic thin layers. In that way, we can separate protons (*incorporated in hydrogenated layer*) from the metal electrode by an intrinsic (*proton free*) layer. The intrinsic layers ($BaTiO_3$) were grown during 15 min under pure argon gas, whereas the hydrogenated layer ($BaTiO_3$:H) was deposited during 90 min under Ar/H_2 gas mixture (80% Ar + 20 % H_2) without breaking the deposition process. **Figure 11** reports the $\varepsilon'(f)$ and $\sigma(f)$ characteristics measured at room temperature on barium titanate films deposited under different atmospheres (**Figure 11(a)**) and in different MIM structures (**Figure 11(b)**).

In **Figure 11(a)** gold was used as metal electrode, and curves labeled (A) were conducted on a sample grown under pure argon gas (*reference*), curves labeled (B) were carried out on a sample grown under 20 % HMR, and curves labeled (C) were done on a tri-layer stack ($BaTiO_3$/$BaTiO_3$:H/$BaTiO_3$). Open symbols denote the dielectric constant ε' and filled symbols represent the conductivity σ. In **Figure 11(b)** measurements were carried out on samples grown under 20 % HMR, where gold [*curves labeled* (B)] and copper [*curves labeled* (D)] were used as metal

http://dx.doi.org/10.5772/65110

electrodes. Further evidences that the observed frequency dispersion of the dielectric constant is related to the electrodes (*and not to the bulk*) are given. At low frequencies, we show that when an intrinsic layer between the hydrogenated bulk film and metallic electrodes was added, both the conductivity and the dielectric constant decrease while remaining higher compared to values measured on Ar-deposited films. This confirms that the low-frequency behavior is related to electrode effects. We previously [25] discussed the dispersive behavior of the $\varepsilon'(f)$ characteristic and attributed it to the proton migration. Protons are indeed very mobile species even at low temperature. Therefore, we can assume that upon the application of a sufficiently high electric field, the positively charged protons move toward the cathode and pile up to form a concentration gradient (*capacitive double layer*). The proton migration leads to an increase of the ionic conductivity and creates a capacitive double layer under the cathode which enhances the dielectric constant. In the case of tri-layer stack, the interfacial Ar-deposited layers react as *dead layers*, which reduce the capacitance of the MIM structure as observed at low frequencies.

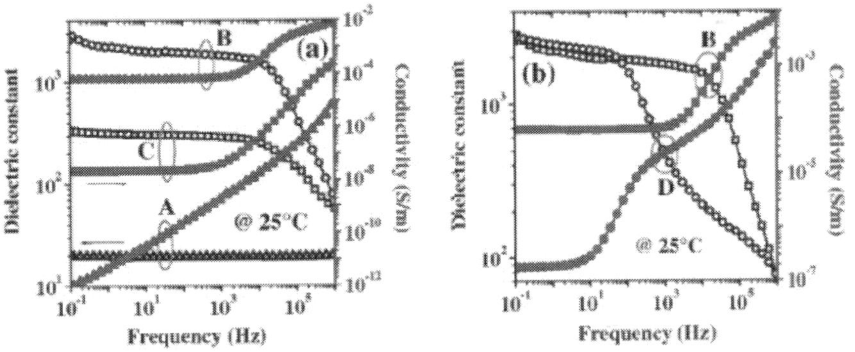

Figure 11. Frequency dependence of the dielectric constant and the conductivity. Measurements were carried out at room temperature on $BaTiO_3$-based MIM capacitors grown under different (a) atmospheres and in different (b) MIM (M = Au, Cu) structures. Curves labeled (A) were conducted on a sample grown under pure argon gas (*reference*), curves labeled (B, D) were carried out on a sample grown under 20 % HMR, and curves labeled (C) were done on a tri-layer stack ($BaTiO_3$/$BaTiO_3$:H/$BaTiO_3$). Open symbols denote the dielectric constant and filled symbols represent the conductivity.

To further check the influence of electrodes, the nature of the metal contact was varied (**Figure 11(b)**). Gold (Au/$BaTiO_3$:H/Au) and copper (Cu/$BaTiO_3$:H/Cu) were used as metal electrodes. It is seen that replacing the Au electrode by Cu is sufficient to modify the conductivity, especially in the low-frequency domain, where the conductivity measured on the Cu/$BaTiO_3$:H/Cu decreases. This clearly shows that the dielectric response is strongly dependent on the electrode nature. Under an electric field, positively charged protons are drifted toward the cathode where, ideally (*if ohmic contact*), they are neutralized by electronic transfer through the electrode/electrolyte interface. In the case of bad charge transfer (*blocking contact*), mobile charges accumulate at the electrode/electrolyte interface, and a space charge region builds up at the electrode boundaries. This feature agrees with decreasing conductivity.

Finally, the last feature which can be observed in **Figure 8** is related to the HT relaxation process ($T > 120$ °C) which can be thermally activated by around 1 eV. This value is close to the usual activation energy reported for oxygen vacancy migration in titanates [43, 44]. These defects, resulting from the deposition under reduced atmosphere, are indeed very mobile species in titanates [43–45]. Therefore, we can assume that upon the application of a low-frequency electric field at sufficiently high temperature, the oxygen vacancies move toward the cathode and pile up at the interface to form a concentration gradient (*space charge*). This feature highly competes with the double layer previously formed by protons which leads to an increase of the electrical conduction with elapsed time.

4. Conclusion

Metal/*a*-$BaTiO_3$:H(1 μm)/metal devices report a specific capacitance around 2 μF/cm^2, steady in the frequency range from 0.1 to 10^4 Hz. This feature is suitable for ac circuits. On the other hand, these devices have low-temperature performance and are capable of delivering energy down to $^-$100 °C with minimal effect on efficiency. Moreover, they can bear high voltages per cell. Compared to conventional capacitor technologies, the studied devices possess orders of magnitude higher energy density. This performance can be rather enhanced by using porous carbon electrodes to achieve a high surface area. In addition, these devices display a low internal resistance, hence providing acceptable power density capability compared to batteries. Finally, the deposition process seems to be compatible with easy incorporation of all-solid-state supercapacitors with other electronic devices. In addition, we can stack supercapacitors vertically to enhance the specific capacitance.

Author details

Fadhel El Kamel

Address all correspondence to: elkf2000@yahoo.fr

LaPhyMNE, University of Gabès, Gabès, Tunisia

References

[1] B. E. Conway. Electrochemical Supercapacitors: Scientific, Fundamentals and Technological Applications. New York: Kluwer Academic/Plenum Publishers, 1999.

[2] L. Ma et al. Appl. Phys. Lett. 2005;87:123503.

[3] M. G. H. Hendriks et al. J. Appl. Phys. 2001;90:5303.

[4] C. C. Liu et al. Electrochim. Acta. 2010;55:5768.

[5] Y. S. Yoon et al. J. Power Sources. 2001;101:126.

[6] J. H. Sung et al. J. Power Sources. 2003;124:343.

[7] C. L. Pint et al. Carbon. 2011;49:4890.

[8] W. Gao et al. Nature Nanotech.2011;6:496.

[9] D. Pech et al. Nature Nanotech. 2010;5:651.

[10] D. Pech et al. J. Power Sources. 2010;195:1266.

[11] W. Sun et al. J. Power Sources. 2010;195:7120.

[12] H. X. Ji et al. Chem. Commun. 2010;46:3881.

[13] J. H. Lim et al. J. Electrochem. Soc. 2001;148:A275.

[14] R. Kötz et al. Electrochim. Acta. 2000;45:2483.

[15] M. Kaempgen et al. Nano. Lett. 2009;9:1872.

[16] J. M. Lee et al. J. Electroceram. 2006;17:639.

[17] J. M. Soon et al. Electrochem. Solid-State Lett. 2007;10:A250.

[18] I. H. Pratt et al. J. Vac. Sci. Technol. 1970;8:256.

[19] S. Furusawa et al. Solid State Ionics. 2005;176:553.

[20] F. El Kamel et al. J. European Ceram. Soc. 2007;27:3807.

[21] F. El Kamel et al. Solid State Ionics. 2009;180:853.

[22] R. Waser. J. Am. Ceram. Soc. 1988;71:58.

[23] J. Ahn et al. Appl. Phys. Lett. 2000;77:1378.

[24] K. Xiong et al. Appl. Phys. Lett. 2004;85:2577.

[25] P. Gonon et al. Appl. Phys. Lett. 2007;90:232902.

[26] P. Gonon et al. J. Appl. Phys. 2007;101:073901.

[27] Y. B. Lin et al. J. Appl. Phys. 2000;87:1841.

[28] S. Zafar et al. Appl. Phys. Lett. 1998;73:175.

[29] C. N. Berglund et al. Phys. Rev. 1967;157:358.

[30] A. J. Moulson and J. M. Herbert. Electroceramics: Materials, Properties and Applications. London: Chapman and Hall; 1996.

[31] D. Keroak et al. J. Phys. C. 1984;17:833.

[32] J. P. Boyeaux et al. J. Phys. C. 1979;12:545.

[33] L. Pejov. Chem. Phys. Lett. 2003;11:376.

[34] L. Dobaczewski et al. Phys. Rev. B. 2004;69:245207.

[35] G. Weber et al. Phys. Rev. B. 1986;34:8406.

[36] J. O. M. Bockris and A. K. N. Reddy. Modern Electrochemistry. New York: Plenum; 1970.

[37] F. El Kamel et al. J. Vac. Sci. Technol. A. 2012;30:04D110-1.

[38] W. Münch et al. Solid State Ionics. 1999;125:39.

[39] E. Matsushita et al. Solid State Ionics. 1997;97:45.

[40] E. Matsushita et al. Solid State Ionics. 1999;125:31.

[41] K. D. Kreuer. Solid State Ionics. 1999;125:285.

[42] K. D. Kreuer. Chem. Mater. 1996;8:610.

[43] F. El Kamel et al. J. Appl. Phys. 2006;99:094107.

[44] R. Waser. J. Am. Ceram. Soc. 1989;72:2234.

[45] R. Waser. J. Am. Ceram. Soc. 1991;74:1934.

Chapter 6

Oxidation-Ultrasound Process on Removing Potassium Ions from Activated Carbon for Improving Electrochemical Properties of Supercapacitor

Kang Sun

Additional information is available at the end of the chapter

http://dx.doi.org/10.5772/64809

Abstract

To produce pure activated carbon (AC) with low potassium ions (K^+) content for supercapacitor, coconut-shell AC activated by KOH was treated with a novel oxidation-ultrasound process on the basis of hydrochloric acid (HCl) washing. The electrochemical performances of the ACs as supercapacitor electrodes were characterized by cyclic voltammetry (CV), galvanostatic charge-discharge (GC), and electrochemical impedance spectroscopy (EIS). Results showed that the obtained AC, which was washed with 1.0 wt% HCl solution for 120 min and subsequently treated with 0.6 wt% H_2O_2 solution at 60°C in an ultrasonic oscillator for 8 h, possessed a K^+ content of 46 mg/kg, much lower than that of 417 mg/kg of the AC without oxidation-ultrasound treatment. Furthermore, a large specific surface area and pore volume of 3460 m^2/g and 1.869 cm^3/g, respectively, were obtained for AC after oxidation-ultrasound treatment. A high specific capacitance of 306 F/g at the current density of 1 A/g in 1 M H_2SO_4 electrolyte was observed for the prepared AC, indicating it has good electrochemical performances, and remained at 294 F/g with a capacitance retention of 96% after 3000 cycles, indicating excellent stability and capacitive behavior of the AC electrode for supercapacitor.

Keywords: activated carbon, potassium ions, oxidation-ultrasound, specific capacitance, supercapacitor

1. Introduction

As a kind of new energy storage device between battery and conventional capacitor, supercapacitor has drawn much attention recently due to its high power density and long cycle life [1–4]. Besides, it was assumed to be one of the most promising energy storage devices for various applications such as brake energy recovery systems, hybrid electrical vehicles, digital telecommunication systems [5, 6], etc. It is known that the surface area, pore size, and purity texture have significant influence on the performances of supercapacitor for the electrode material [7]. So far, AC has been widely applied in electrode material for supercapacitor applications, owing to its advantages of good chemical stability and conductivity, large surface area, low cost [8–11], etc. However, the AC products show a high K^+ content due to the process limitations of KOH activation, as a result, the already-formed AC pores were blocked, which had a negative effect on the specific capacitance and cycle life of supercapacitor. Therefore, it is necessary to remove the K^+ from the AC, aiming to greatly improve the electrochemical performances of supercapacitor.

In general, the K^+ existed in the AC often spreads in the pores, and HCl washing cannot efficiently remove it. Although ultrasound oscillation was effective to disperse K^+ into solution, oxidation process is able to weaken the force between AC pore walls and K^+, resulting in deeply purification of the AC samples. Therefore, this study aims to make use of a novel oxidation-ultrasound process to reduce the K^+ content of ACs. The effect of K^+ content on AC properties will be reflected by the physical and electrochemical performances.

2. Experimental

2.1. Preparation of AC sample

In this study, coconut shell was used to prepare AC by KOH activation. First of all, the coconut shell was subjected to be carbonized at 600°C for 1 h and was crushed to obtain a particle size of about 1 mm. Then 50 wt% KOH solution with the KOH/C weight ratio of 4:1 was mixed with the coconut shell carbide in a stainless steel reactor. In addition, the mixture was activated in a muffle furnace in which the mixture was first pretreated at 350°C for 2 h under air atmosphere, then heated up to 800°C at a heating rate of 10°C min^{-1}, and maintained at 800°C for 1 h under sealed condition. After cooled down to the room temperature, the resultant samples were subsequently washed with distilled water until a pH of approximately 6 was obtained. Finally, the prepared AC samples were dried at 120°C for 10 h under vacuum condition.

2.2. Removal of the K+ from AC sample

There are two steps included in the removal process of the K^+. In the first step, 2.0 g as-obtained AC was impregnated with 100 ml 0.3, 0.5, 0.7, 1.0, and 1.2 wt% of HCl solution, respectively. Then the mixture was stirred at 80°C for 30–150 min in an electro-thermostatic water cabinet. The oxidation-ultrasound process was used to purify the HCl-washed AC deeply in the second step. In an ultrasonic oscillator, 100 ml different weight ratios 0, 0.2, 0.4, 0.6, 0.8, and 1.0 wt% of H_2O_2 solution were mixed with 2.0 g HCl-washed AC for 2–10 h at 10–80°C. The final treated samples were dried at 120°C under vacuum condition overnight. After that, the resulting AC samples were named AC-W, AC-W-H, and AC-W-H-U, corresponding to the water, HCl, and oxidation-ultrasound treatment on AC, respectively.

2.3. Characterizations

Inductively coupled plasma mass spectrometry (ICP-MS) was used to measure the K^+ content of ACs. Scanning electron microscopy (SEM, S-3400, Hitachi) was used to exhibit the surface morphologies of the samples. Nitrogen (77 K) adsorption was carried out using a Micromeritics ASAP 2020 analyzer to detect the porous properties of ACs. The specific surface area values, total pore volume, and the pore size distribution of activated carbons were calculated by the BET method, according to the adsorbed amount and density functional theory (DFT), respectively.

2.4. Electrochemical characterization

The capacitive performance measurements were carried out at room temperature in a three-electrode system on an electrochemical workstation (Bio-Logic, France) in 1 M H_2SO_4 electrolyte. The active AC materials, acetylene black, and 5 wt% PTFE with a weight ratio of 8.5:1:0.5 were fully mixed and grounded in a mortar. Then the mixture was pressed onto a Ni-foam current collector to form the working electrode. And Ag/AgCl and platinum foil were employed as the reference and counter electrode, respectively. The capacitive behavior of the AC electrode was characterized by CV, GCD, and electrochemical impedance spectroscopy (EIS) measurements. CV was performed from 0.01 to 0.9 V at various scan rates in a range of 5–200 mV s^{-1}, GC curve was measured from 0.01 to 0.9 V at different current densities ranging from 0.1 to 5 Ag^{-1}, and EIS was recorded at the frequency from 100 kHz to 10 MHz.

3. Results and discussion

3.1. Influencing factors on removal process of K^+

3.1.1. Effect of HCl treatment on the K^+ content of AC

In this paper, HCl solution is first used to treat with the AC-W samples to remove the K^+. The effects of the treatment time and mass fraction of HCl solution are investigated in detail. It can be seen in **Figure 1(a)** that the relationship between the mass fraction of HCl solution and K^+ content for 90 min. And the figure also shows the K^+ content exhibits a conspicuous decrease with the increasing mass fraction of HCl solution from 0 to 1.0 wt%, and then it presents a rising trend. The excessive acid radical ions results in K^+ removed from AC difficulty during electric double layer process, which is formed on the surface of K^+. Thus, the optimal mass fraction of HCl solution is 1.0 wt%.

It shows the relationship between K^+ content and the treatment time of HCl solution at the mass fraction of 1.0 wt% in **Figure 1(b)**. The content of K^+ decreases from 1242 to 417 mg/kg significantly, while the treatment time increases from 30 min to 2 h, and it does not change anymore after 2 h. Therefore, treatment time of 120 min and 1.0 wt% HCl solution are chosen as the optimal treatment conditions for the whole experiments, and the corresponding AC sample is named AC-W-$H_{1\%/120min}$.

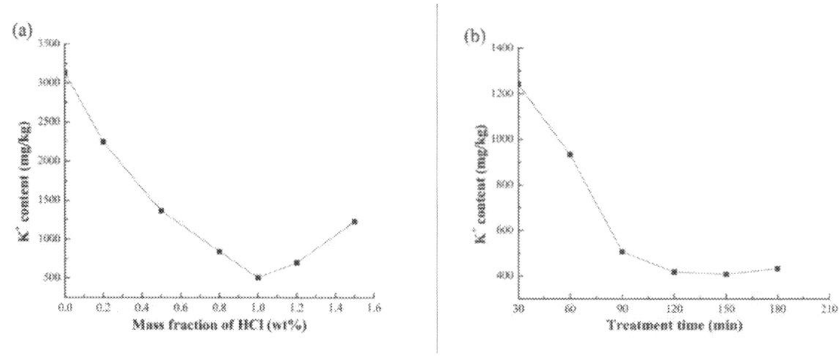

Figure 1. Effect of mass fraction and treatment time of HCl solution on the K^+ content of AC.

3.1.2. Effect of oxidation-ultrasound process on the K^+ content of AC

The AC-W-$H_{1\%/120min}$ is then subjected to oxidation-ultrasound process to deeply remove the K^+. The influence of the mass fraction of H_2O_2 solution, ultrasound temperature, and ultrasound time on the K^+ content of ACs is studied. The obtained results are shown in **Table 1** and **Figure 2**. The mass fraction of H_2O_2 solution is in the range of 0–1.0 wt% at an ultrasound time of 6 h and an ultrasound temperature of 60°C. **Table 1** indicates that the K^+ content of AC without H_2O_2 treatment is 256 mg/kg, substantially lower than that of the AC-W-$H_{1\%/120min}$ with 417 mg/kg, which should have close relationships with the ultrasound effect that the ultrasonic wave can accelerate the mass transfer process of them between the solid-liquid phase and promote some uncarbonized substances inside the AC pores to be dispersed into the solution along with the K^+. It can be seen from **Figure 2(a)** that the K^+ content of AC decreases markedly from 256 to 61 mg/kg, and thereafter it remains substantially unchanged when the mass fraction of H_2O_2 solution increased from 0 to 0.6 wt%. Therefore, it can be stated that H_2O_2 treatment shows a great effect on producing ACs with low K^+content. On the one hand, as an oxidizing agent, H_2O_2 solution can reduce the force between the K^+ and AC pore walls, leading to it transferred to the solution easily. On the other hand, the K^+ in the AC can coexist with some organic groups in the form of chemical bond, which can be destroyed by the presence of H_2O_2 solution, making it possible for the K^+ of ACs to be removed effectively [12, 13].

Mass fraction of H_2O_2/wt%	0	0.2	0.4	0.6	0.8	1.0
K^+ content/(mg/kg)	256	184	92	61	64	75
Ultrasonic temperature/°C	10	20	40	60	70	80
K^+ content/(mg/kg)	157	124	96	61	74	91
Ultrasonic time/h	2	4	6	8	10	12
K^+ content/(mg/kg)	148	119	61	46	41	44

Table 1. Results of mass fraction of H_2O_2 solution, ultrasound temperature and ultrasound time on the K^+ content of AC.

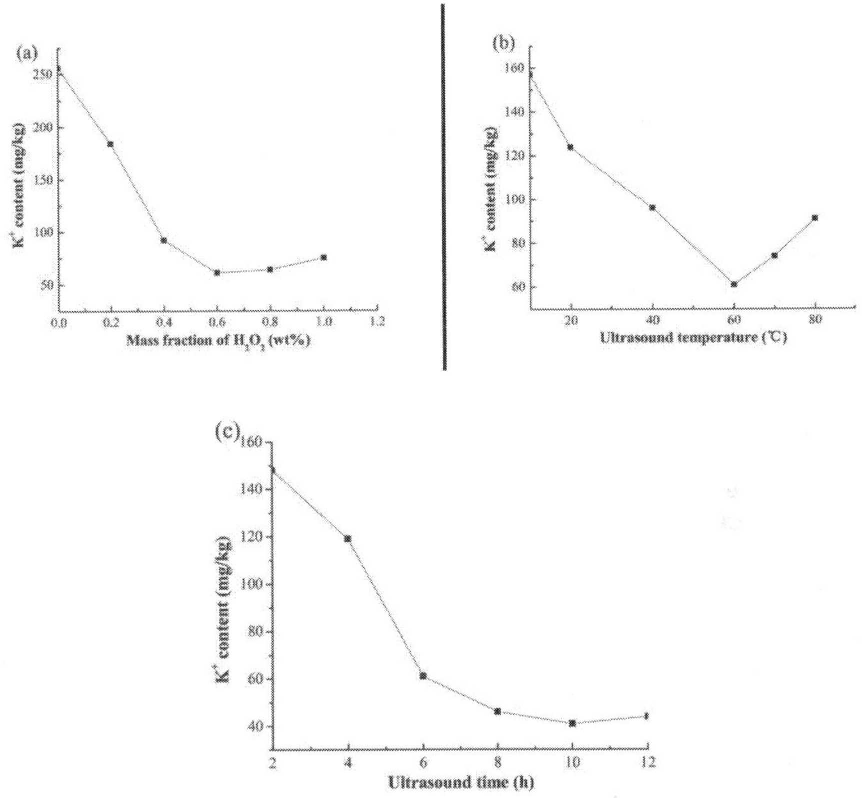

Figure 2. Effect of mass fraction of (a) H_2O_2 solution, (b) ultrasound temperature and (c) ultrasound time on the K^+ content of AC.

Figure 2(b) exhibits the relationship between K^+ content and the ultrasound temperature in the range of 10~80°C at an ultrasound time of 6 h, using 0.6 wt% H_2O_2 solution. It was observed that the K^+ content of AC decreased from 157 to 61 mg/kg when the temperature increased from 10 to 60°C, but increased while the temperature was further increased from 60 to 80°C. This might be attributed to the fact that the energy and activity of the K^+ were improved with the increase of temperature in ultrasonic field and the movement of K^+ into the solution was accelerated. **Figure 2(c)** shows the effect of ultrasound time on the K^+ content of AC, using 0.6 wt% H_2O_2 solution at the ultrasound temperature of 60°C. It is clear that the K^+ content drops from 148 to 46 mg/kg as the ultrasound time increases from 2 to 8 h, and it scarcely changes with a continuing increase of ultrasound time. Therefore, we consider 0.6 wt% H_2O_2 solution, ultrasound temperature of 60°C and ultrasound time of 8 h as the optimum experiment conditions to deeply remove the K^+ from AC, and the corresponding AC is marked as $AC\text{-}W\text{-}H_{1\%/120min}\text{-}U_{0.6\%/60°C/8h}$.

3.2. Sample characterization

3.2.1. Morphology characterization

Figure 3(a–c) shows the surface morphologies of the AC-W, AC-W-$H_{1\%/120min}$, and AC-W-$H_{1\%/120min}$-$U_{0.6\%/60°C/8h}$ by SEM at 1000× magnification. It can be seen clearly that the morphologies of the AC samples display an obvious change on the surface. The AC-W has large quantities of uncarbonized substances presented on the surface. The AC-W-$H_{1\%/120min}$ has relatively flat and smooth surface compared to the AC-W, but there are still some residues on the surface. Meanwhile, as shown in **Figure 3(c)**, the surface of the AC-W-$H_{1\%/120min}$-$U_{0.6\%/60°C/8h}$ was very clean and smooth, confirming that the oxidation-ultrasound process has a significant effect on the purification of AC, which is consistent with the aforementioned results of the K^+ content.

Figure 3. SEM images of (a) AC-W, (b) AC-W-$H_{1\%/120min}$ and (c) AC-W-$H_{1\%/120min}$-$U_{0.6\%/60°C/8h}$.

3.2.2. Porous texture characterization

It was reported that the specific surface area and pore size of AC were critical factors for electrochemical capacitor applications [14], hence, the N_2 adsorption-desorption isotherms and corresponding pore size distribution curves of the as-prepared ACs were analyzed as shown in **Figure 4**. **Table 2** gives the textural characteristics and specific capacitance of the samples. It was noticed that the presented isotherms were type I for all samples according to IUPAC classification [15, 16]. The N_2 adsorption volume of the AC-W achieved saturation at low relative pressure of 0.1 and the lowest isotherm was observed, suggesting that a great number of activating agents (KOH) are still blocking in the AC pores, especially the mesopores. The AC-W-$H_{1\%/120min}$ displays significantly higher isotherm than the AC-W and its N_2 adsorption volume shows a gradual increase until the relative pressure of 0.4, implying that most of the KOH remained in the AC pores are removed by HCl washing. As for the AC-W-$H_{1\%/120min}$-$U_{0.6\%/60°C/8h}$. The existence of a hysteresis loop in the desorption branch at the relative pressure of 0.47 was observed obviously. And the isotherm with a highest adsorbed volume indicated that the deep removal of the K+ along with other residues from the AC pores by the oxidation-ultrasound process greatly increased the mesopore volumes.

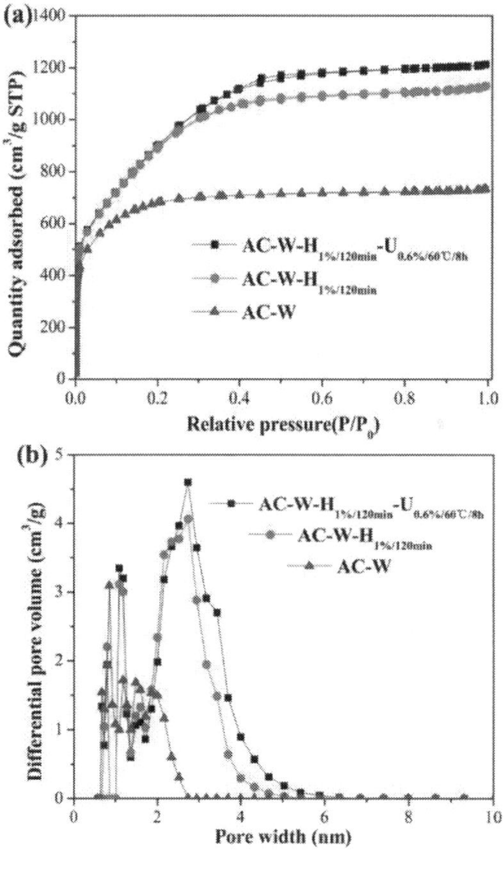

Figure 4. (a) N_2 adsorption-desorption isotherms and (b) the corresponding pore size distributions of the as-prepared samples.

Samples	S_{BET} (m^2/g)	V_t (cm^3/g)	V_{meso} (cm^3/g)	V_{micro} (cm^3/g)	D_p (nm)	C_g (F/g)	Ash (wt.%)	Yield (%)
AC-W	2198	1.110	0.311	0.799	1.781	147	1.2	—
AC-W-$H_{1\%/120min}$	3178	1.686	1.297	0.389	2.086	255	0.1	94.3
AC-W-$H_{1\%/120min}$-$U_{0.6\%/60°C/8h}$	3460	1.869	1.615	0.254	2.228	306	0.01	97.5

Note: S_{BET}: specific surface area; V_t: total pore volume; V_{meso}: mesopore volume; V_{micro}: micropore volume; D_p: average pore size; C_g: specific gravimetric capacitance; yield: the independent result of every step for removing K^+.

Table 2. Textural characteristics and specific capacitance of the as-prepared samples.

Figure 4(b) gives the pore size distributions of the as-prepared ACs. Results showed that the pore size of AC-W was in the range of 0.6–2 nm, while that of AC-W-$H_{1\%/120min}$-$U_{0.6\%/60°C/8h}$ and AC-W-$H_{1\%/120min}$ ranged from 2 to 4 nm and 0.5 to 2.0 nm, respectively. It is obvious that the purification of the AC-W with HCl and oxidation-ultrasound treatment in turn gradually made mesopores (2–4 nm) become the dominant type of pores.

It can be seen from **Table 2** that the AC-W had an average pore size of 1.781 nm a mesopore volume of 0.311 cm^3/g, while the AC-W-$H_{1\%/120min}$ and AC-W-$H_{1\%/120min}$-$U_{0.6\%/60°C/8h}$ achieve the mesopore volume of 1.297 and 1.615 cm^3/g, and the average pore size of 2.086 and 2.228 nm, respectively. Moreover, the AC-W-$H_{1\%/120min}$ has significantly larger surface area and total pore volume than the AC-W, increasing respectively from 2198 to 3178 m^2/g and 1.110 to 1.686 cm^3/g, and the AC-W-$H_{1\%/120min}$-$U_{0.6\%/60°C/8h}$ possesses the largest surface area and pore volume of 3460 m^2/g and 1.869 cm^3/g. It is evident that the removal of K^+ from AC samples creates some new microspores and obviously promotes the widening of existing micropores into mesopores at the same time, which can be accessible to electrolyte ions for electrical double-layer formation [17]. Thus, the large surface area, high pore volume, and low K^+ content of AC samples make them good candidates for electrode materials.

3.3. Electrochemical characterization

CV is used in determination of the electrochemical performances of as-prepared samples. The typical CV results of AC-W, AC-W-$H_{1\%/120min}$ and AC-W-$H_{1\%/120min}$-$U_{0.6\%/60°C/8h}$ at the scan rate of 10 mV/s with the potential range of −0.2~0.8 V are shown in **Figure 5(a)**. Results showed that all the AC electrodes displayed an approximately quasi-rectangular voltammogram shape, which was the characteristic of electrochemical double-layer capacitance [18–20]. Furthermore, the CV curve of AC-W-$H_{1\%/120min}$-$U_{0.6\%/60°C/8h}$ electrode exhibits a bigger current response and larger area of rectangle than that of the AC-W and AC-W-$H_{1\%/120min}$ electrode, demonstrating an obvious increase in capacity during the removal process of K^+. **Figure 5(b)** shows the CV curves of the AC-W-$H_{1\%/120min}$-$U_{0.6\%/60°C/8h}$ electrode at different scan rates of 5–50 mV/s. The CV curve eventually becomes titled with the increase of the scan rate, but still maintains a rectangular-like shape even at 50 mV/s, implying a small resistance in the accessible pores and an excellent capacitive behavior [21, 22] of the AC-W-$H_{1\%/120min}$-$U_{0.6\%/60°C/8h}$ electrode.

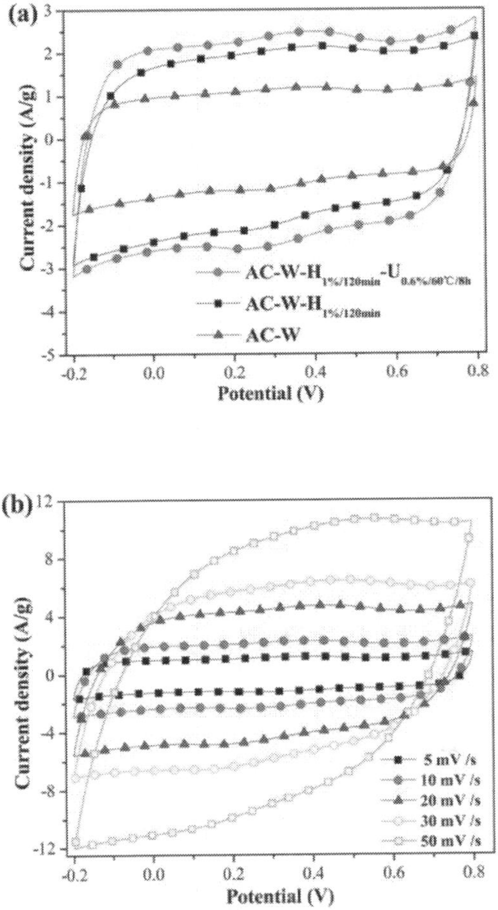

Figure 5. Cyclic voltammograms of the samples: (a) all AC electrodes at a scan rate of 10 mV/s and (b) AC-W-$H_{1\%/120min}$-$U_{0.6\%/60°C/8h}$ with different scan rates.

GC measurements have also been conducted to investigate the electrochemical performances of as-prepared samples. **Figure 6(a)** shows the GC curves of the AC-W, AC-W-$H_{1\%/120min}$ and AC-W-$H_{1\%/120min}$-$U_{0.6\%/60°C/8h}$ at the current density of 1 A/g with the potential range of −0.2–0.8 V. It can be seen that all the samples present a virtually linear shape and isosceles triangle curve, indicating a good reversibility and typically capacitive behavior of the AC electrodes [23, 24]. According to the GC curves, the AC-W-$H_{1\%/120min}$-$U_{0.6\%/60°C/8h}$ has longer discharging time compared to the AC-W and AC-W-$H_{1\%/120min}$, implying larger specific capacitance, which proved that the deep removal of K^+ from AC samples had a significant effect on the electrochemical performance of AC electrode.

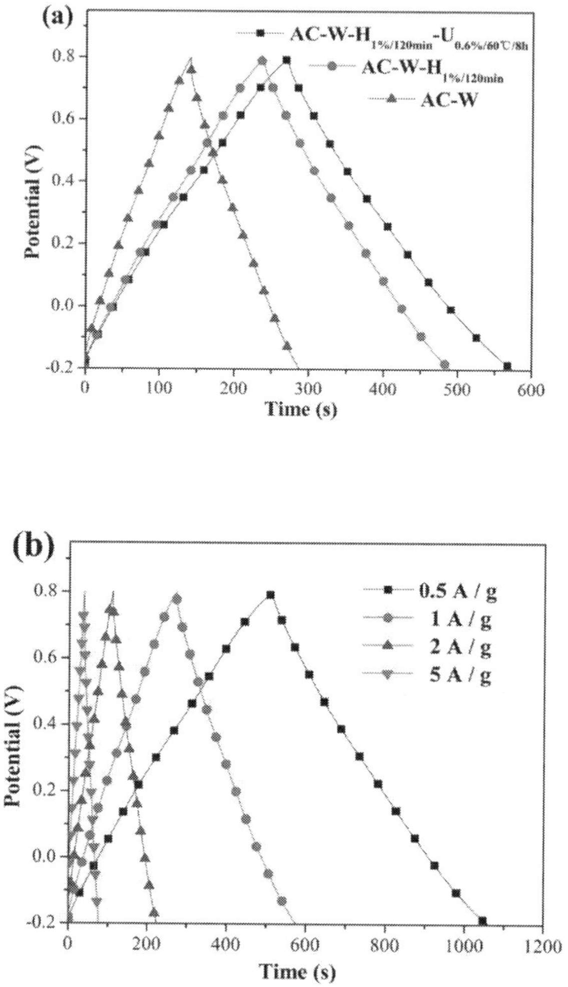

Figure 6. Galvanostatic charge/discharge curves of the samples: (a) all activated carbon electrodes at a current density of 1 A/g and (b) AC-W-$H_{1\%/120min}$-$U_{0.6\%/60°C/8h}$ with different current densities.

Figure 6(b) represents the GC curves of the AC-W-$H_{1\%/120min}$-$U_{0.6\%/60°C/8h}$ measured in a current density range of 0.5–5 A/g. It was obvious that the discharging time dropped as the current density increased, while the curve was constant as typical triangle shape even at a high loading current density of 5 A/g, revealing that the AC-W-$H_{1\%/120min}$-$U_{0.6\%/60°C/8h}$ as electrode material is promising for a high performance supercapacitor. In addition, the specific capacitance of the AC electrodes can be calculated from the charge-discharge curves based on the following equation [25, 26]:

$$C_g = 2\frac{I \cdot \Delta t}{m \cdot \Delta V} \tag{1}$$

where C_g is the specific gravimetric capacitance (F/g), I is the current loaded, Δt is the discharge time (s), ΔV is the potential change during the discharge process, and m (g) represents the mass of the AC. As shown in **Table 2**, the AC-W-$H_{1\%/120min}$-$U_{0.6\%/60°C/8h}$ has the highest specific gravimetric capacitance of 306 F/g, which is more than twice as large as that of the AC-W (147 F/g), and increases by about 20% compared with the AC-W-$H_{1\%/120min}$ (255 F/g). The enhanced specific capacitance of AC samples can be ascribed to the enhancement of the surface area, effective pore volume and purity, as confirmed by the above results of K^+ content and N_2 adsorption-desorption isotherms.

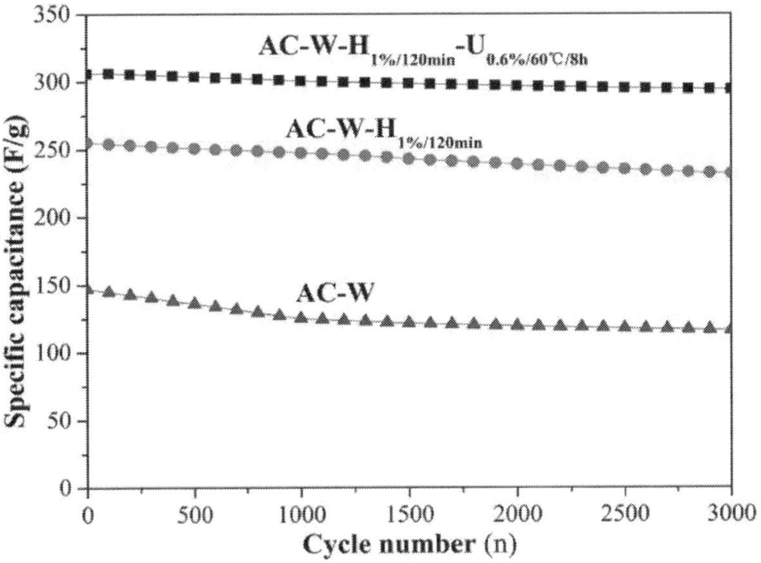

Figure 7. Cyclic performances of the samples at a current density of 1 A/g.

Figure 7 shows the cyclic stability of AC-W, AC-W-$H_{1\%/120min}$ and AC-W-$H_{1\%/120min}$-$U_{0.6\%/60°C/8h}$ electrodes detected by galvanostatic charge-discharge at the current density of 1 A/g for 3000 cycles. The specific capacitance of AC-W decreased obviously from 147 to 125 F/g after 1000 cycles and was maintained at about 116 F/g with the capacitance retention of 79% after 3000 cycles. The specific capacitance of AC-W-$H_{1\%/120min}$ is relatively stable and achieves capacitance retention of 91% after 3000 cycles. The specific capacitance for AC-W-$H_{1\%/120min}$-$U_{0.6\%/60°C/8h}$ is found to be 306 F/g in the first cycle and 294 F/g after 3000 cycles with a coulombic efficiency of 96%, revealing the excellent stability and reversibility of the AC-W-$H_{1\%/120min}$-$U_{0.6\%/60°C/8h}$ electrode.

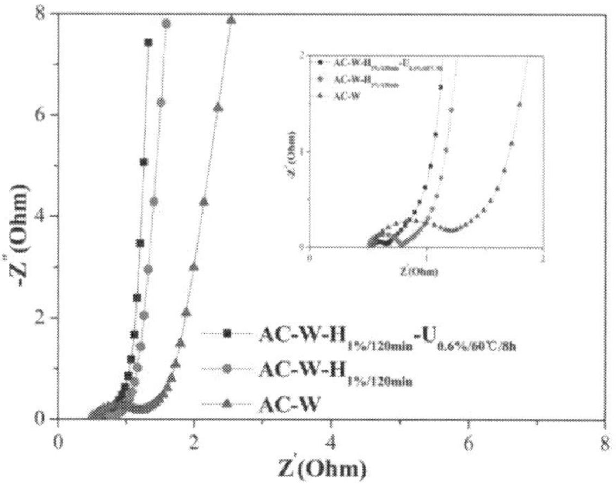

Figure 8. Nyquist plots of the as-prepared samples (inset: enlarged high-frequency region of Nyquist plots).

EIS measurements are employed to obtain impedance performance information. Typical Nyquist plots for AC-W, AC-W-$H_{1\%/120min}$ and AC-W-$H_{1\%/120min}$-$U_{0.6\%/60°C/8h}$ electrodes are presented in **Figure 8**. Results revealed that a semicircle in the high-frequency region and a straight line in the low-frequency region were observed in all the three plots [27]. And the plot of AC-W-$H_{1\%/120min}$-$U_{0.6\%/60°C/8h}$ sample displays a relatively smaller semicircle than that of AC-W and AC- W-$H_{1\%/120min}$ samples in the high-frequency region, indicating a lower charge transfer resistance. This may be ascribed to the larger mesopore volume and lower K^+ content of AC-W-$H_{1\%/120min}$-$U_{0.6\%/60°C/8h}$, which can facilitate the rapid diffusion of electrolyte ions into the pores of electrode materials. In addition, the AC-W-$H_{1\%/120min}$-$U_{0.6\%/60°C/8h}$ shows more vertical line leaning to imaginary axis in the low-frequency region [28], suggesting better capacitive behavior than AC-W and AC-W-$H_{1\%/120min}$.

4. Conclusions

In this work, ACs with high performance for supercapacitor are prepared from coconut shell by KOH activation, using a novel oxidation-ultrasound process to deeply remove the K^+ of AC. The experiment results demonstrate that the AC samples washed with 1.0 wt% HCl solution for 120 min and subsequently treated with 0.6 wt% H_2O_2 solution at 60°C in an ultrasonic oscillator for 8 h possess a very low K^+ content of 46 mg/kg. Compared to the AC-W and AC-W-$H_{1\%/120min}$, the AC-W-$H_{1\%/120min}$-$U_{0.6\%/60°C/8h}$ exhibits a larger surface area and pore volume of 3460 m^2/g and 1.869 cm^3/g. As the electrode material for electrochemical application, the AC-W-$H_{1\%/120min}$-$U_{0.6\%/60°C/8h}$ showed a high specific capacitance of 306 F/g with a coulombic efficiency of 96% after 3000 cycles. The oxidation-ultrasound process has a great potential to produce ACs with high performance for supercapacitor applications.

http://dx.doi.org/10.5772/64809

Acknowledgements

This research is financially supported by the projects in Forestry Public Benefit Research Sector (Grant No. 201404610) and the National Natural Science Foundation of China (Grant No. 31400510).

Author details

Kang Sun

Address all correspondence to: sunkang0226@163.com

Institute of Chemical Industry of Forest Products, CAF, Nanjing, China

References

[1] X.J. He, Y.J. Geng, J.S. Qiu, M.D. Zheng, S. Long and X.Y. Zhang, Carbon, 2010, 48, 1662-1669.

[2] B.E. Conway, Electrochemical supercapacitors. Scientific fundamentals and technological applications. Kluwer Academic Plenum Press, New York, 1999.

[3] X.L. Li, C.L. Han, X.Y. Chen and C.W. Shi, Microporous and Mesoporous Materials, 2010, 131, 303-309.

[4] T.E. Rufford, D. Hulicova-Jurcakova, K. Khosla, Z.H. Zhu and G.Q. Lu, Journal of Power Sources, 2010, 195, 912-918.

[5] H. Gómez, M.K. Ram, F. Alvi, P. Villalba, E.L. Stefanakos and A. Kumar, Journal of Power Sources, 2011, 196, 4102-4108.

[6] C.S. Yang, Y.S. Jang and H.K. Jeong, Current Applied Physics, 2014, 14, 1616-1620.

[7] F. Gao, G.H. Shao, J.Y. Qu, S.Y. Lv, Y.Q. Li and M.B. Wu, Electrochimica Acta, 2015, 155, 201-208.

[8] S. Kumagai, M. Sato and D. Tashima, Electrochimica Acta, 2013, 114, 617-626.

[9] T. Zhang, M.B. Zheng, N.W. Li, H.L. Lu, S.T. Zhang and J.M. Cao, Materials Letters, 2013, 105, 43-46.

[10] V. Ruiz, C. Blanco, E. Raymundo-Piñero, V. Khomenko, F. Béguin and R. Santamaría, Electrochimica Acta, 2007, 52, 4969-4973.

[11] H.R. Yu, S. Cho, M.J. Jung and Y.S. Lee, Microporous and Mesoporous Materials, 2013, 172, 131-135.

[12] J.J. Rueda-Márquez, M. Sillanpää, P. Pocostales, A. Acevedo and M.A. Manzano, Water Research, 2015, 71, 85-96.

[13] Y.W. Wang, Y.S. Wei and J.X. Liu, Journal of Hazardous Materials, 2009, 169, 680-684.

[14] Y. Xiao, C. Long, M.T. Zheng, H.W. Dong, B.F. Lei, H.R. Zhang and Y.L. Liu, Chinese Chemical Letters, 2014, 25, 865-868.

[15] X.J. He, P.H. Ling, M.X. Yu, X.T. Wang, X.Y. Zhang and M.D. Zheng, Electrochimica Acta, 2013, 105, 635-641.

[16] S.J. Han, Y.H. Kim, K.S. Kim and S.J. Park, Current Applied Physics, 2012, 12, 1039-1044.

[17] N.D. Kim, W. Kim, J.B. Joo, S. Oh, P. Kim, Y. Kim and J. Yi, Journal of Power Sources, 2008, 180, 671-675.

[18] L. Wei and G. Yushin, Carbon, 2011, 49, 4830-4838.

[19] M.X. Wang, C.Y. Wang, M.M. Chen, Y.S. Wang, Z.Q. Shi, X. Du, T.Q. Li and Z.J. Hu, New Carbon Materials, 2010, 25, 285-290.

[20] X. Du, C.Y. Wang, M.M. Chen, Y. Jiao and J. Wang, Journal of Physical Chemistry C, 2009, 113, 2643-2646.

[21] H.J. Shen, E.H. Liu, X.X. Xiang, Z.Z. Huang, Y.Y. Tian, Y.H. Wu, Z.L. Wu and H. Xie, Materials Research Bulletin, 2012, 47, 662-666.

[22] M.B. Wu, R.C. Li, X.J. He, H.B. Zhang, W.B. Sui and M.H. Tan, New Carbon Materials, 2015, 30, 86-91.

[23] X.X. Xiang, E.H. Liu, Z.Z. Huang, H.J. Shen, Y.Y. Tian, C.Y. Xiao, J.J. Yang and Z.H. Mao, Journal of Solid State Electrochemistry, 2011, 15, 2667-2674.

[24] D.S. Patil, S.A. Pawar, R.S. Devan, Y.R. Ma, W.R. Bae, J.H. Kim and P.S. Patil, Materials Letters, 2014, 117, 248-251.

[25] A. Davies and A. Yu, Canadian Journal of Chemical Engineering, 2011, 89, 1342-1357.

[26] M.B. Wu, P.P. Ai, M.H. Tan, B. Jiang, Y.P. Li, J.T. Zhang, W.T. Wu, Z.T. Li, Q.H. Zhang and X.J. He, Chemical Engineering Journal, 2014, 245, 166-172.

[27] J.L. Chang, Z.Y. Gao, X.R. Wang, D.P. Wu, F. Xu, X. Wang, Y.M. Guo and K. Jiang, Electrochimica Acta, 2015, 157, 290-298.

[28] I. Oh, M. Kim and J. Kim, Microelectronics Reliability, 2015, 55, 114-122.

Section 2

Supercapacitor Applications

Chapter 7

Development and On-Orbit Demonstration of Lithium-Ion Capacitor-Based Power System for Small Spacecraft

Masatoshi Uno and Akio Kukita

Additional information is available at the end of the chapter

http://dx.doi.org/10.5772/64966

Abstract

Lithium-ion capacitors (LICs) offer higher energy density and specific energy than do traditional electric double-layer capacitors (EDLCs). In spacecraft power systems where traditional lithium-ion batteries (LIBs) have been used with shallow depth of discharge (DoD) in order to achieve long-cycle life, LICs would potentially be an alternative to secondary batteries. Firstly, this chapter presents the quantitative comparison between the LIB- and LIC-based spacecraft power system from the viewpoint of system mass. On the basis of the potential suggested by the comparison, we have been developing the technology demonstration platform named "NESSIE" that contains the LIC pouch cell as one of its major demonstration missions. NESSIE was successfully launched with the main satellite HISAKI on September 2013. This chapter also presents the development of the LIC pouch cell for NESSIE and its experimental (or ground test) and on-orbit operation data.

Keywords: cycle life testing, lithium-ion capacitor, pouch cell, spacecraft power system, vacuum tolerance

1. Introduction

Applications of secondary battery-based energy storage are rapidly expanding from portable electronic devices to large-scale systems, including electric vehicles and grid-connected applications. Among various secondary battery chemistries, lithium-ion batteries (LIBs) are the most promising and viable for portable and vehicular applications thanks to their highest energy density and specific energy. Vigorous research and development efforts for increased energy density and extended service life are underway.

Meanwhile, supercapacitors (SCs), formally known as electric double-layer capacitors (EDLCs), are also an attractive energy storage device that plays an important role in various scenes. Thanks to their energy storage mechanism utilizing the double-layer capacitance on porous-activated carbon electrodes, EDLCs offer remarkable advantages over traditional secondary batteries in terms of cycle life, power capability, safety, and allowable temperature range. However, their low-energy density property is generally considered as a major drawback to be used as an alternative energy storage source to traditional secondary batteries. Hence, applications of EDLCs are chiefly limited to hybrid energy storage applications where EDLCs play a role of the high-power energy buffer that complements the main energy sources, such as secondary batteries and fuel cells.

To cope with the drawback of the low-energy density of EDLCs, lithium-ion capacitors (LICs) have been developed and commercialized by several manufacturers. LICs are basically a hybrid electrochemical capacitor combining double-layer capacitance and lithium intercalation for energy storage mechanism, and they offer the higher specific energy than EDLC without sacrificing the major benefits of EDLCs. LICs have been drawing significant attentions in vehicular and industry applications where high-power energy buffers are indispensable. Despite the increased energy density property, there is still a huge gap between LICs and secondary batteries.

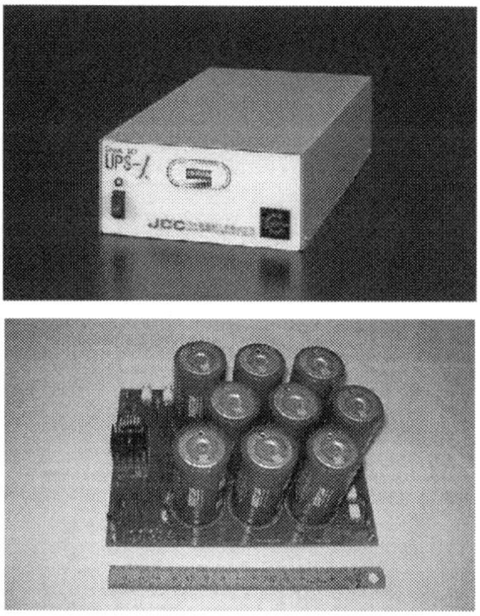

Figure 1. EDLC-based uninterruptible power supplies (UPSs).

However, once the excellent long-cycle life performance of EDLCs and LICs is factored in, it drives expectation that EDLCs and LICs would be an alternative energy storage source to traditional secondary batteries in certain applications. For example, EDLC-based uninterruptible power supplies (UPSs) have been commercialized [1], as shown in **Figure 1**, as maintenance-free alternatives aiming for infrastructure applications where traditional lead-acid batteries need to be replaced with new ones for every few years. Another potential application is a spacecraft power system where long-life energy storage sources are essential. In the previous study [2, 3], systematic cycle life testing emulating low-Earth-orbit (LEO) satellite conditions has been performed for EDLCs and LICs, and the results demonstrated that excellent cycle life performance can be expected for cycling conditions for LEO satellites.

For spacecraft applications where specific energy is of great importance, LICs are undoubtedly superior to EDLCs. Despite the great potential suggested in the laboratory cycle life testing, LICs had not been used in practical space applications. In order to demonstrate the LIC performance in practical on-orbit conditions, we developed the technology demonstration platform named 'NESSIE (NExt-generation Small Satellite Instrument for Electric power systems)' containing an LIC pouch cell as one of its major components to be demonstrated on orbit; NESSIE's major missions are (1) on-orbit demonstration for the LIC cells and (2) on-orbit demonstration for high-efficiency thin-film solar cells. NESSIE was embedded in the main satellite named "HISAKI", the Spectroscopic Planet Observatory for Recognition of Interaction of Atmosphere (see **Figure 2**) and was launched on September 14, 2013.

Figure 2. An image of HISAKI.

In this chapter, the quantitative comparison between LIB- and LIC-based spacecraft power systems is performed in terms of system mass. This chapter also presents the development of the LIC pouch cell for NESSIE and its experimental (or ground test) and on-orbit operation data.

2. Cycle life of LICs

In the previous study [4], cycle life testing emulating LEO spacecraft conditions was performed for LICs, and capacitance retentions of LICs were compared to those of LIBs under the same temperature conditions, as shown in **Figure 3**. The capacitance retention of the LIB with the shallow DoD of 20% dropped to approximately 70% at 10,000 cycles, which is equivalent to approximately 2-year operation, while that of LICs was still greater than 95%.

Furthermore, different from LIBs whose retention trends were dependent on DoD, the LICs showed DoD-independent trends.

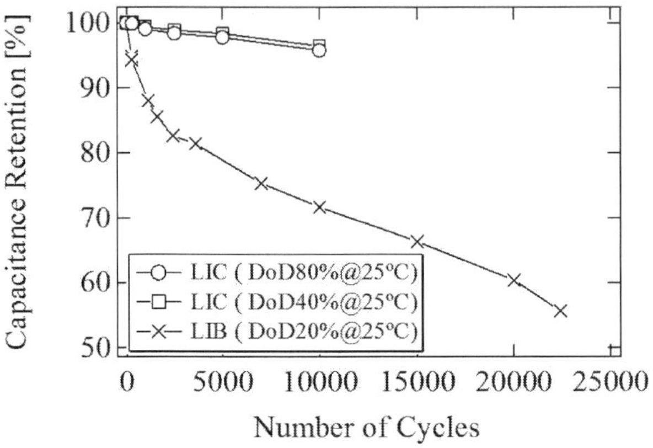

Figure 3. Capacitance retention trends of LIBs and LICs [4].

The accelerated ageing testing and the cycle life prediction model for LICs were also investigated in the previous study [3]. LIC cells were cycled at various temperatures in the systematically designed cycle test matrix. The activation energies of degradation ratios of LIC cells were calculated using the Arrhenius equation, whereupon ageing acceleration factors were determined. The experimental and predicted capacitance retention trends matched satisfactorily, and hence, the established cycle life prediction model was verified. According to the reported results in [3], LICs are expected to operate even longer than millions of cycles at temperature lower than 10°C.

The mission life of the main satellite HISAKI was 1 year, and therefore, the cycle life of LICs reported in the previous studies was adequate for NESSIE. However, the cycle life testing reported in the previous study was performed under laboratory conditions, leaving the investigation of tolerance against space environment, such as vacuum, vibration, and radiation, as issues to be cleared. Section 4 discusses these issues in detail.

3. LIC- and LIB-based power system comparison

3.1. Specific energy

Cycle life performance of traditional secondary batteries is greatly dependent on DoD. In general, secondary batteries for LEO satellites are operated with DoD shallower than 40% in order to meet the typical cycle life requirement of 30,000 cycles that is equivalent to approximately 5.7 years on-orbit operation. In other words, the net specific energy of secondary batteries is far lower than their rated specific energy because of the shallow DoD operation. Specific energies as well as net specific energies of various energy storage devices are compared in **Table 1**. Although LIBs offer the highest specific energy of 150 Wh/kg, their net specific energy drops to as low as 60 Wh/kg with DoD of <40%.

	Secondary battery		**Capacitor**	
	Alkaline battery (Ni-Cd, Ni-MH, and Ni-H_2)	**Lithium-ion battery (LIB)**	**Electric double-layer capacitor (EDLC)**	**Lithium-ion capacitor (LIC)**
Specific energy	40–60 Wh/kg	150 Wh/kg	<10 Wh/kg	<30 Wh/kg
Depth of discharge	<40%	<40%	<80%	<80%
Net specific energy	<24 Wh/kg	<60 Wh/kg	<8 Wh/kg	<24 Wh/kg

Table 1. Specific energy and net specific energy of energy storage devices.

EDLCs and LICs, on the other hand, offer excellent cycle life performance that is insensitive to DoD conditions, allowing deep DoD operations. In the case of the 80% DoD operation for EDLCs and LICs, for example, their net specific energies are <8 Wh/kg and <24 Wh/kg, respectively, greatly bridging the gap in terms of net specific energy.

3.2. Photovoltaic array reduction by constant power charging scheme

A charging profiles of a 10-Wh LIB cell with a constant-current-constant-voltage (CC-CV) charging scheme are shown in **Figure 4(a)** as typical characteristics. The charging power peaks at the end of the CC charging period (or at the beginning of the CV charging period), when both the current and voltage become maximum. The chargeable state of charge (SoC) in the CC charging period is merely 60%, and the rest 40% is charged with CV charging. This characteristic indicates that the CV charging is indispensable for LIBs to reach high SoC. In typical LEO satellites, photovoltaic (PV) arrays are designed to be capable of supplying not only load power but also this peak charging power.

Different from LIBs mentioned above, LICs and EDLCs can be charged to nearly 100% SoC without CV charging because of their low-impedance properties. A charging profiles of a 10-Wh LIC cell with the constant-power (CP) charging scheme, instead of CC-CV charging scheme, are shown in **Figure 4(b)**. The charging power with the CP charging scheme is constant, and its peak value is substantially lower than that with the CC-CV charging scheme. This reduction in peak charging power contributes to the reduction in area and mass of PV arrays, as will be exemplified in Section 3.6.

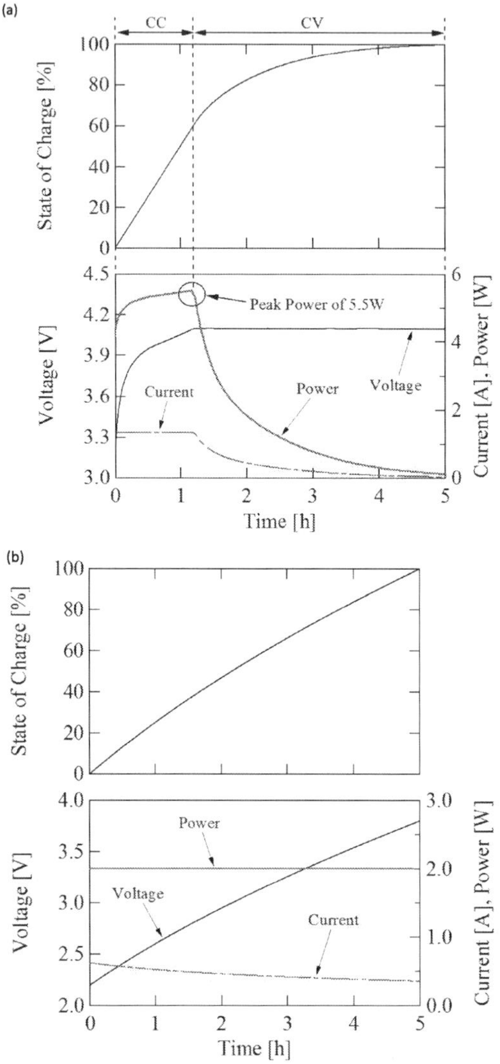

Figure 4. Comparison between (a) CC-CV charging and (b) CP charging schemes.

3.3. Power system using LIC

An unregulated bus system using an array power regulator (APR), as shown in **Figure 5(a)**, is mainstream for low- to medium-scale spacecraft power systems using an LIB-based energy storage. The LIB can be directory connected to the load (or bus) because of the relatively float voltage characteristics. The APR plays a role of battery charging, and a load power is supplied by the LIB and/or APR.

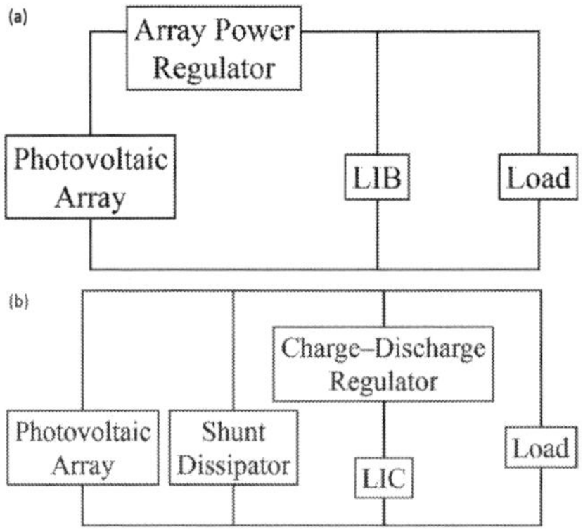

Figure 5. Spacecraft power system architectures for (a) LIB- and (b) LIC-based systems.

On the other hand, an LIC cannot be directory connected to the bus because of the relatively large-voltage variation (see **Figure 4(b)**), and therefore, a charge-discharge regulator (CDR) is necessary. An LIC-based power system using a shunt dissipator as a bus voltage regulator is shown in **Figure 5(b)**. In the following subsections, power systems shown in **Figure 5** are compared from the viewpoint of mass.

3.4. Mass of LIB-based power system

A single charge-discharge cycle consists of the sun and eclipse periods T_{sun} and $T_{eclipse}$, respectively. The discharged energy during the eclipse period E_{dis_LIB} is

$$E_{dis\_LIB} = P_{load} T_{eclipse} \tag{1}$$

where P_{load} is the load power.

Batteries contain not only cells but also mechanical structural support that represents nearly 20% increase in mass over that of cells. Now, the ratio of the mechanical support to cells is defined as A. The mass of the LIB, M_{LIB}, is expressed as

$$M_{LIB} = \frac{E_{dis\_LIB}(1+A)}{DS_{LIB\_cell}} \tag{2}$$

where D is the DoD, and S_{LIB_cell} is the specific energy of LIB cells.

Let V_{ave} be the average voltage of the LIB during discharging. The capacity of the LIB, C_{LIB}, can be determined as

$$C_{LIB} = \frac{E_{dis\_LIB}}{DV_{ave}} \tag{3}$$

Introducing the charge rate (also known as C-rate) as R_{cha}, the charge current for the LIB, I_{cha}, is yielded as

$$I_{cha} = C_{LIB} R_{cha} \tag{4}$$

The value of R_{cha} can be determined from the charge-discharge ratio that is defined as

$$R_{C/D} = \frac{R_{cha} T_{sun}}{D} \tag{5}$$

Generally, a typical value of $R_{C/D}$ for LEO spacecraft is 1.25 [5].

The maximum charging power (P_{cha_LIB}) at the end of the CC charging or at the beginning of the CV charging, at which the LIB voltage is as high as V_{cha}, is

$$P_{cha\_LIB} = V_{cha} I_{cha} \tag{6}$$

The mass of the PV arrays in the LIB-based power system, M_{PV_LIB}, can be determined to be

$$M_{PV\_LIB} = \frac{P_{load} + P_{cha\_LIB}}{\eta_{APR} \rho_{PV}} \tag{7}$$

where η_{APR} is the power conversion efficiency of the APR and ρ_{PV} is the specific power of the PV array.

The mass of the power conditioning system M_{PCS_LIB}, which is equal to the mass of the APR in the LIB-based power system, is expressed as

$$M_{PCS\_LIB} = m_{APR} \left(P_{cha\_LIB} + P_{load} \right) \tag{8}$$

where m_{APR} is the mass/watt coefficient of the APR.

From Eqs. (2), (7), and (8), the total mass of the LIB-based power system M_{LIB_system} is obtained as

$$M_{LIB\_system} = M_{LIB} + M_{PV\_LIB} + M_{PCS\_LIB} \tag{9}$$

3.5. Mass of LIC-based power system

Let η_{CDR} be the power conversion efficiency of the CDR. The discharged energy of the LIC E_{dis_LIC} can be expressed as

$$E_{dis\_LIC} = \frac{P_{load} T_{eclipse}}{\eta_{CDR}} \tag{10}$$

With the specific energy of the LIC cells S_{LIC_cell}, the mass of the LIC, M_{LIC}, can be represented as

$$M_{LIC} = \frac{E_{dis\_LIC}(1 + A)}{DS_{LIC\_cell}} \tag{11}$$

Assuming that the LIC is fully charged at the end of the sun period, the power demanded for the PV array from the LIC, P_{cha_LIC}, is expressed as

$$P_{cha\_LIC} = \frac{E_{dis\_LIC}}{\eta_{CDR} T_{sun}} \tag{12}$$

The mass of the PV arrays in the LIC-based power system, M_{PV_LIC}, is

$$M_{PV\_LIC} = \frac{P_{load} + P_{cha\_LIC}}{\rho_{PV}} \tag{13}$$

Assuming the shunt dissipator is designed to be capable of the sum of P_{load} and P_{cha_LIC}, the mass of the PCS in the LIC-based power system, M_{PCS_LIC}, is given by

$$M_{PCS\_LIC} = P_{load} m_{CDR} + m_{shunt} \left(P_{cha\_LIC} + P_{load} \right) \tag{14}$$

where m_{CDR} and m_{shunt} are the mass/watt coefficient of the CDR and shunt dissipator, respectively.

From Eqs. (11), (13), and (14), the total mass of the LIC-based power system M_{LIC_system} is determined as

$$M_{LIC\_system} = M_{LIC} + M_{PV\_LIC} + M_{PCS\_LIC} \tag{15}$$

3.6. System mass comparison

The LIB- and LIC-based power systems, shown in **Figure 5**, are quantitatively compared using the parameters listed in **Table 2**. These values were determined according to the literature [6].

Parameter	Symbol	Value
Load power	P_{load}	1000 W
Sun period	T_{sun}	1 h
Eclipse period	$T_{eclipse}$	0.5 h
C/D ratio (for LIB only)	$R_{C/D}$	1.25
Charge voltage (LIB only)	V_{cha}	28.7 V
Average discharge voltage (LIB only)	V_{ave}	25.9 V
Mass ratio of mechanical structural supports to cells	A	20%
Specific power of PV array	ρ_{PV}	60 W/kg
Efficiency of array power regulator	η_{APR}	90%
Efficiency of charge/discharge regulator	η_{CDR}	90%
Mass/watt coefficient of array power regulator	m_{APR}	4 kg/kW
Mass/watt coefficient of charge/discharge regulator	m_{CDR}	4 kg/kW
Mass/watt coefficient of shunt dissipator	m_{Shunt}	2 kg/kW

Table 2. Parameters used for mass comparison.

Figure 6 depicts the system mass of 1-kW LIB- and LIC-based power systems, with nominal specific energies (S_{LIB_cell} = 150 Wh/kg and S_{LIC_cell} = 30 Wh/kg) ±20% variation, as a function of DoD. As mentioned earlier, in order for LIBs to meet the typical cycle life requirement of 30,000 cycles, DoD should be set as shallow as 40% to mitigate cycling-induced degradation. The mass of the LIB-based power system with 40% DoD is around 50 kg and lighter than the LIC-based system, indicating the traditional LIB-based power system is superior for the typical 30,000-cycle life requirement from the viewpoint of system mass. For longer cycle life requirement, on the other hand, LIBs have to be cycled with even shallower DoD, and therefore, the mass of the LIB-based power system is prone to sharp increase, as can be found around 20% DoD in **Figure 6**. Hence, the LIB-based system with shallow DoD competes with the LIC-based system. The mass of the LIB-based system with DoD of 15%, for example, is around 70 kg and is comparable with that of the LIC-based system with 60% DoD.

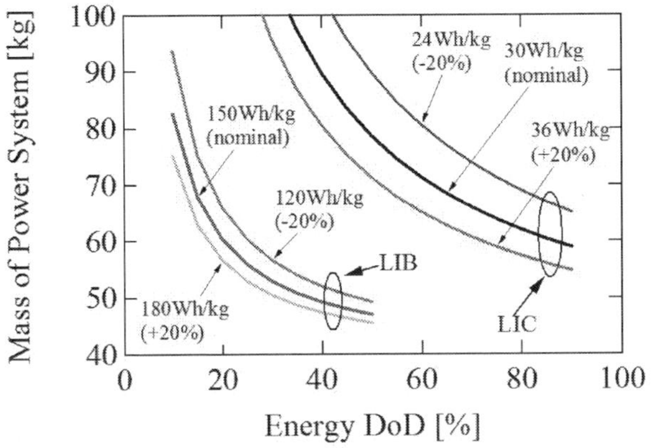

Figure 6. Mass of LIB- and LIC-based power systems as a function of DoD.

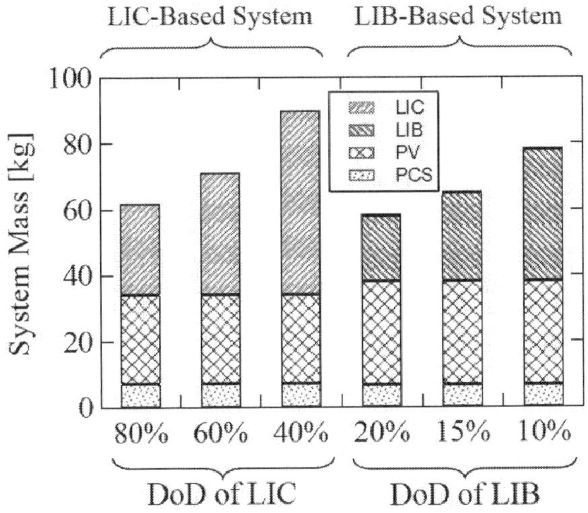

Figure 7. Mass breakdown of LIB- and LIC-based power systems.

The mass breakdown of the LIB- and LIC-based power systems with the nominal specific energies of S_{LIB_cell} = 150 Wh/kg and S_{LIC_cell} = 30 Wh/kg are compared in **Figure 7**. In the LIC-based power systems, the LIC accounted for greater than half the total mass of the systems. Meanwhile, thanks to the CP charging scheme introduced in Section 3.2, the mass of the PV array in the LIC-based system can be reduced compared to that in the LIB-based system. The total mass of the LIC-based system with DoD of 60–80% is comparable with that of LIB-based system with DoD of 15–20%, suggesting that LICs would be an alternative energy storage source to LIBs for applications needing long-cycle life. Other benefits, such as wider operational temperature range of LICs, would further improve the likelihood of LICs being an alternative to traditional LIBs.

4. Development of LIC for NESSIE

4.1. Vacuum tolerance of LIC pouch cell

Needless to say, there is no air in the space, and therefore, components for spacecraft must be vacuum-tolerant. In general, secondary batteries for spacecraft power systems are reinforced by metal-housing so as to increase the ruggedness for shock and vibration during launch and tolerance against vacuum in space. In the previous developmental work, electrical characteristics of lithium-ion pouch cells were investigated [7]. The cells swelled in vacuum, and the performance of LIB pouch cells significantly deteriorated. In order to improve the vacuum tolerance, LIB pouch cells were potted with epoxy resin in an aluminium-housing. However, this reinforcement adversely increases the mass and volume of the battery, resulting in decreased specific energy, and it neutralizes the benefit of pouch cells of high specific energy.

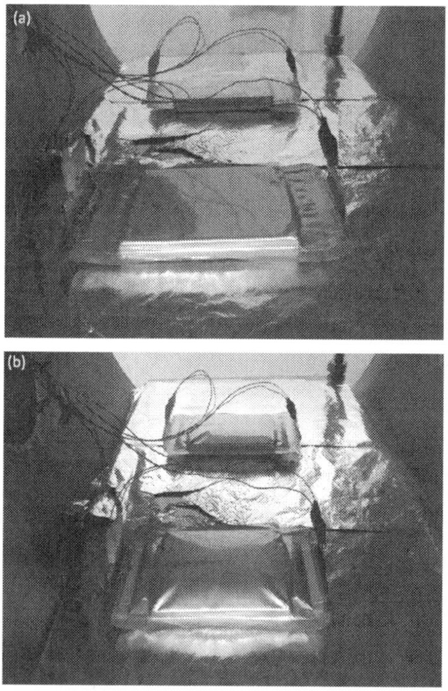

Figure 8. LIC pouch cells under (a) normal pressure and (b) vacuum.

Although both the resin- and metal-housing-based reinforcement were reportedly necessary for LIB pouch cells, we considered it was worth investigating whether LIC pouch cells work well in vacuum. To investigate vacuum tolerance of LIC pouch cells, we performed short-term cycle life testing for two LIC pouch cells in a vacuum chamber. In the vacuum chamber, LIC pouch cells were placed on a thermostatic plate that was controlled to be 25°C by a coolant circulator (see **Figure 8**). As reference data, the same cycling test under a normal pressure condition was also carried out for other two LIC pouch cells. A single charge-discharge cycle consists of 65-min CP charging and 35-min CC discharging with 80% DoD.

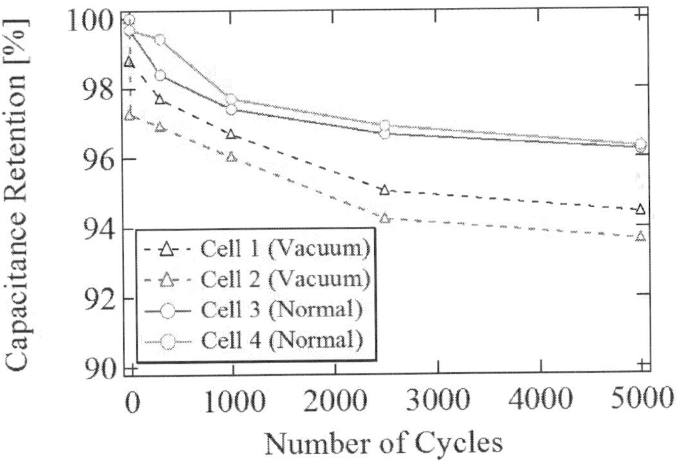

Figure 9. Capacitance retention trends of LICs cycled under normal pressure and vacuum conditions.

The pouch cells under normal pressure and vacuum are shown in **Figure 8(a)** and **(b)**, respectively. The LIC pouch cells swelled similarly to the LIB pouch cells reported in [7]. In spite of the significant shape deformation, the observed deterioration of cells in a vacuum was merely around 2%, compared to those under normal pressure, as shown in **Figure 9**. The capacitance retentions of cells dropped as the vacuum chamber was evacuated at the beginning of the testing (i.e., at an initial cycle). After this initial drop in capacitance retention, all cells exhibited the similar degradation trends. The measured capacitance retentions of cells in vacuum and under normal pressure at 5000th cycle (equivalent to 1-year on-orbit operation) were approximately 94 and 96%, respectively. **Figure 10** shows the discharge curve trends of LICs during capacitance measurement. The dischargeable time gradually shortened as cells deteriorated. However, no significant voltage decline due to an increase in internal resistance was observed, indicating that the vacuum condition did not increase an internal resistance of LIC pouch cells. The results shown in **Figures 9** and **10** suggested that bulky and heavy reinforcement using resin and/or metal-housing would not be mandatory even though pouch cells swelled.

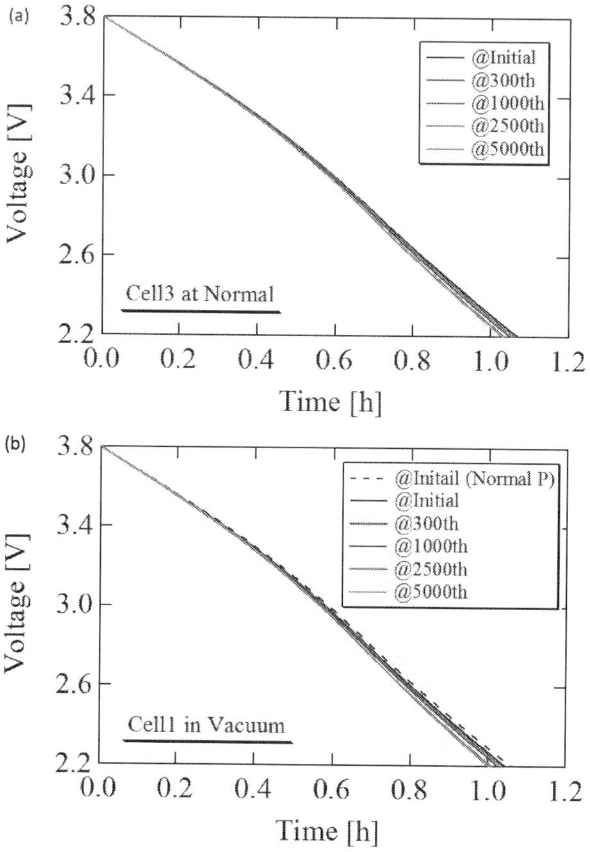

Figure 10. Discharge curve trends of LICs cycled under (a) normal pressure and (b) vacuum.

4.2. LIC pouch cell for NESSIE

NESSIE is a small demonstration platform, in which only one single LIC cell is allowed to be equipped. In general, for practical use, a plastic or metal container is used for pouch cells to be stacked and bundled to form a module. However, to reduce the mass of NESSIE, a container for the LIC pouch cell should be as simple and light as possible. In addition, although the results shown in **Figures 9** and **10** implied that LIC cells might be used without reinforcement, the swelling should be avoided because it might cause unexpected interference with other components in NESSIE. To prevent the swelling at the lightest possible measure, we developed a dedicated metal-bracket that looks 'III-shape', as shown in **Figure 11**. This metal bracket can suppress the swelling at a light mass. This LIC pouch cell with the metal bracket successfully passed the vibration test. The specification of the LIC cell for NESSIE is shown in **Table 3**.

Figure 11. LIC cell with metal bracket.

Mass	303 g (bare cell) 832 g (w/metal bracket)
Dimension	123 × 165 × 15 mm (bare cell) 164 × 220 × 23 mm (w/metal bracket)
Manufacture	Asahikasei FDK energy device
Capacitance	2500 F

Table 3. Specification of LIC cell.

5. Development of NESSIE

5.1. Charge-discharge regulator for LIC

The LIC cell is charged and discharged by a CDR that is basically a bidirectional dc-dc converter. As explained in the previous section, LIC cells for the short-term cycle life testing in the vacuum chamber were charged and discharged with the CP charging and CC discharging schemes, respectively. In practical use, the bidirectional dc-dc converter plays the role of these charging and discharging schemes.

In general, a discharging power is simply determined by loads, and the bidirectional converter operates to regulate a load voltage at a constant value (e.g., 5 V in the NESSIE's power system). On the other hand, a charging power is controlled by the bidirectional converter based on measured voltage and current. To this end, a feedback control loop including current and voltage sensors is necessary. Furthermore, the measured current and voltage values need to be multiplied to determine the charging power. In general, for small-scale systems such as NESSIE, both circuit- and system-level simplifications are of great importance to realize the miniaturized circuit and system. In other words, the current and voltage sensors as well as computational circuit for determining the charging power are desirably be eliminated from the system.

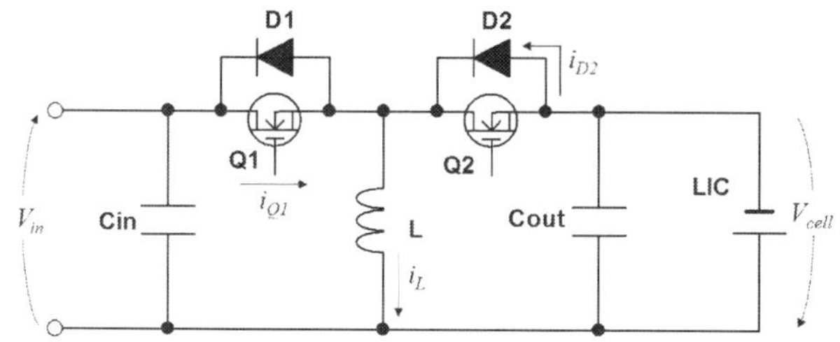

Figure 12. Bidirectional buck-boost converter as charge-discharge regulator for LIC.

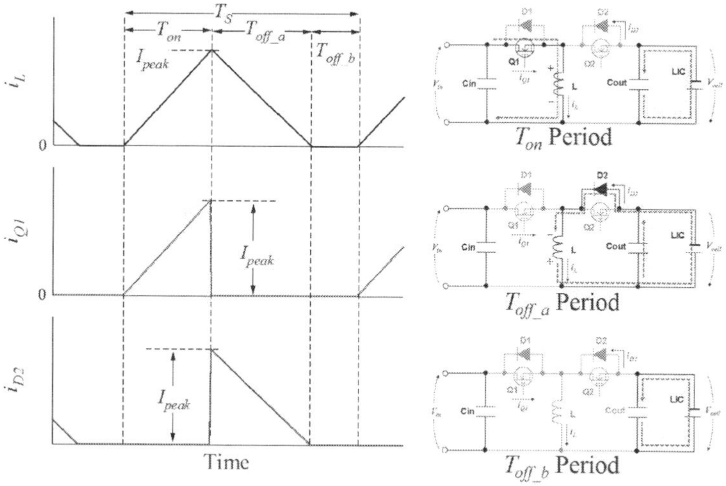

Figure 13. Key operation waveforms and current flow directions in DCM operation.

As a simplest possible solution, we employed a bidirectional buck-boost converter operating in the discontinuous conduction mode (DCM), with which the charging power can be automatically constant even without feedback control nor calculation of product of current and voltage. The bidirectional buck-boost converter and its key operation waveforms as well as current flow directions in DCM are shown in **Figures 12** and **13**, respectively. Switches Q_1 and Q_2 are driven so that an LIC is charged and discharged, respectively. The fundamental operation of the buck-boost converter is well known and can be found in basic textbooks, and hence, this subsection focuses on the mechanism of the CP charging. One switching cycle contains three modes, and T_{off_b} period, during which no current flows in the converter, is unique to DCM operation; the inductor current i_L reaches zero in the T_{off_b} period for every switching cycle. In this chapter, the operation for charging only is explained to save page length. To charge an LIC with the buck-boost converter shown in **Figure 12**, the switch Q_1 is driven while Q_2 is always off. For discharging, on the other hand, the operation is vice versa —Q_2 is driven, and Q_1 is always off.

In the first mode, T_{on} period, i_L increases linearly from zero, and its inclination is equal to V_{in}/L. At the end of this mode, i_L reaches its peak value of I_{peak} expressed as

$$I_{peak} = \frac{V_{in} T_{on}}{L} = \frac{V_{in} D T_S}{L} \tag{16}$$

where T_S is the switching period and D is the duty cycle defined as T_{on}/T_S.

As Q_1 is turned off, the operation moves to T_{off_a} period, during which i_L linearly decreases from I_{peak} with the slope of $-V_{cell}/L$. i_L starts flowing through diode D_2 that is connected in parallel with Q_2. The time length of this period, T_{off_a}, can be determined to be

$$T_{off\_a} = I_{peak} \frac{L}{V_{cell}} = \frac{V_{in} D T_S}{V_{cell}} \tag{17}$$

As i_L reaches zero, the final mode of T_{off_b} period begins. In this mode, except for the current of the smoothing capacitor C_{out}, no current flows in the converter.

In order for T_{off_b} period to exist, T_{off_a} must be shorter than T_S - T_{on}, meaning $T_{off\_a} < T_S - T_{on}$. It leads to the operation criterion given by

$$D < \frac{V_{cell}}{V_{in} + V_{cell}} \tag{18}$$

The input current for the buck-boost converter is supplied only during T_{on} period, and hence, the input power (or charging power) can be expressed as

$$P_{in} = V_{in} \frac{I_{peak}}{2} \frac{T_{on}}{T_S} \tag{19}$$

This equation indicates that, with a constant on-time of T_{on}, the charging power for the LIC cell or P_{in} for the buck-boost converter can be automatically constant even without feedback control. In other words, open-loop control is feasible, hence allowing simplified control circuit.

The photograph of a power control unit (PCU) that contains the buck-boost converter is shown in **Figure 14**. The converter operated with a fixed D of 0.35 at a switching frequency of 100 kHz for the charging power to be approximately 3.8 W. For discharging periods, on the other hand, D was adjusted so that the load voltage was controlled to be 5.0 V.

Figure 14. Photograph of power control unit (PCU) containing charge-discharge regulator.

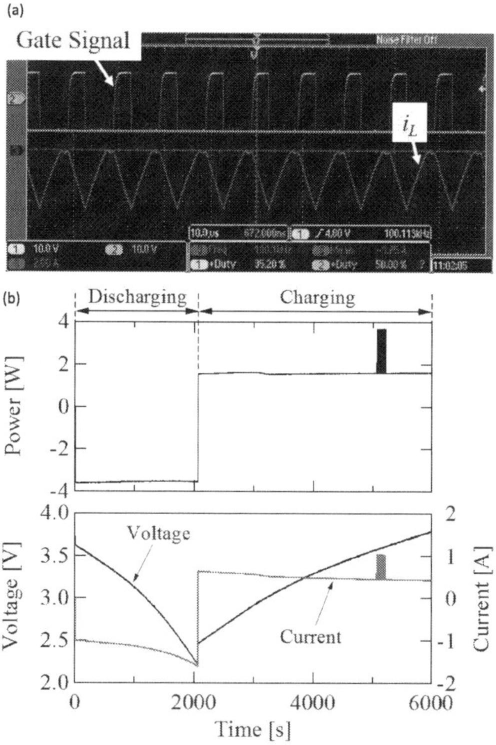

Figure 15. (a) Measured key operation waveforms of the buck-boost converter and (b) cycling profiles of an LIC cycled by the buck-boost converter.

The measured key operation waveforms during charging are shown in **Figure 15(a)**. The inductor current i_L swung as the gate signal for Q_1 was applied. The measured i_L was discontinuous triangular wave—the displayed i_L is inverted—and the good agreement with the theoretical ones was observed. A single charge-discharge cycling profile of an LIC cell with the buck-boost converter operating in DCM is shown in **Figure 15(b)**. The measured power during charging was nearly constant, verifying the automatic CP charging property.

5.2. NESSIE

The specification of NESSIE is listed in **Table 4**. The photographs of NESSIE installed in the main satellite HISAKI are shown in **Figure 16**. The high-efficiency thin-film solar cells were mounted on the main panel that was used as a lid for the interior components, including the LIC pouch cell and charge-discharge regulator. NESSIE was installed in the side panel of the main satellite HISAKI.

Mass	10.03 kg
Dimension	550 × 463 × 205 mm
Bus	5-V fully regulated
Power generation	>10 W (at end of life)

Table 4. Specification of NESSIE.

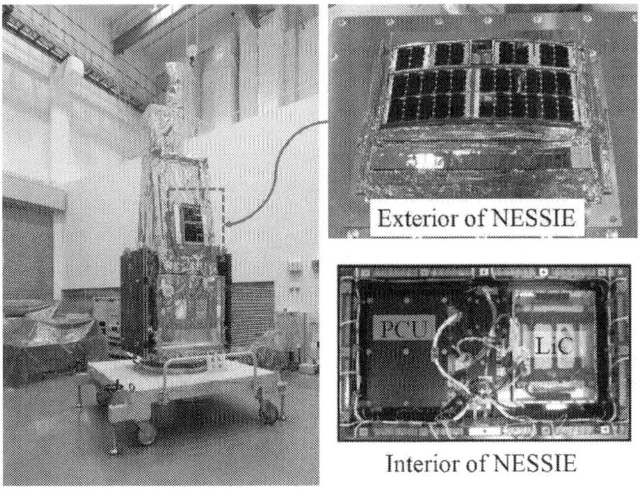

Figure 16. Photographs of NESSIE installed in HISAKI.

6. On-orbit operation

The main satellite HISAKI was launched by Epsilon-1 rocket on September 14, 2013, and NESSIE was turned on for the first time on October 12, 2013, and the LIC was cycled in the first check-up operation, as shown in **Figure 17**. During the first eclipse period, the LIC was nearly fully discharged to 2.2 V, followed by a CP charging during the subsequent sun period. The data at the beginning of the charging were temporarily not available because of an operation with low communication rate, with which essential housekeeping data for HISAKI only was acquired. The short-term pulsating discharges during charging periods were due to the measurement operation for the thin-film solar cells, during which the PV panel was short-circuited and the LIC supplied power to the loads. The first check-up results demonstrated the LIC as well as the charge-discharge regulator performed well.

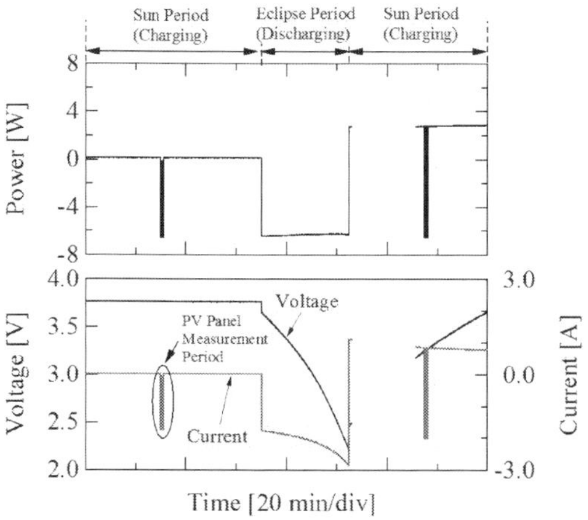

Figure 17. On-orbit cycling profiles during the first check-up.

Since approximately 1 month after the first check-up, the voltage of the PV panel of NESSIE has unexpectedly dropped. The decreased PV panel voltage was lower than the threshold voltage of the charge-discharge regulator. In other words, the voltage of the PV panel was not high enough for the charge-discharge regulator to operate, meaning the LIC has no longer been charged since then.

As the best effort we could do with the malfunctioning PV panel, we determined to turn on NESSIE only for a few minutes in every few months to observe the LIC pouch cell under the float condition. This is rather different from the originally planned cycling condition that was used for laboratory testing. But we consider that this floating operation of the LIC pouch cell would still be meaningful to demonstrate the on-orbit vacuum tolerance and to investigate whether unexpected malfunctions would happen to the LIC after long-term exposure to space environment.

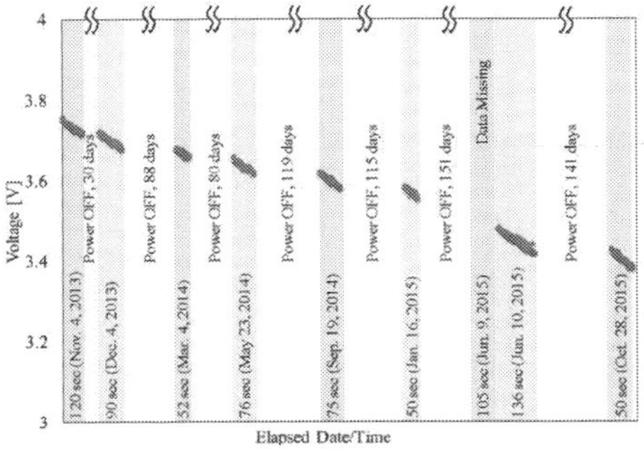

Figure 18. On-orbit trend of LIC voltage.

The trend of the LIC voltage is shown in **Figure 18**. As aforementioned, NESSIE was turned on for a few minutes in every few months, and the LIC-powered NESSIE and its voltage gradually decreased during the turn-on periods. During the turn-off periods, on the other hand, the voltage of the LIC unchanged as the voltages at the beginning and end of turn-off periods were nearly identical, indicating the insignificant self-discharge.

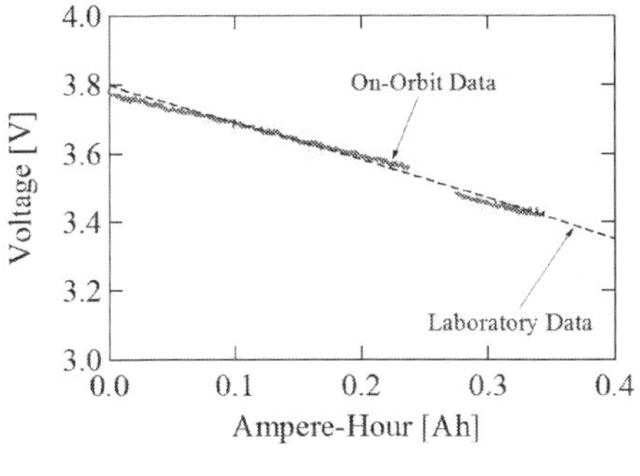

Figure 19. On-orbit trend of LIC voltage as a function of discharged ampere-hour capacity.

The on-orbit LIC's voltage trend shown in **Figure 18** is redrawn as a function of discharged capacity in ampere-hour and is compared to that of a cell tested in the laboratory, as shown in **Figure 19**. The on-orbit and laboratory data matched very well, suggesting the LIC has been performing well.

7. Conclusions

Although the specific energy of LICs is far lower than that of traditional LIBs, the gap between LICs and LIBs can be greatly bridged once long-cycle life performance with deep

DoD conditions is factored in. In addition, LICs allow the CP charging scheme, with which the mass of the PV panel can be reduced compared to that with a traditional CC-CV charging scheme. The mass of an LIC-based power system was compared to that of a LIB-based system. The results of the quantitative comparison suggested that, for applications requiring very long-cycle life, for which LIBs need to be cycled with DoD shallower than 20% to achieve long-cycle life, the LIC-based power system would be comparable to the LIB-based one, driving expectation that LIC would potentially be an alternative energy storage source.

We developed the technology demonstration platform named NESSIE whose one of the missions is the on-orbit demonstration of the LIC pouch cell. Short-term cycle life testing in vacuum was performed to investigate the vacuum tolerance of LIC pouch cells. The results of the 1-year testing suggested the deterioration in capacitance retention due to the vacuum condition was insignificant. The dedicated charge-discharge regulator, with which the LIC can be charged with the CP charging scheme even without feedback control, was also developed to realize simplified circuit.

NESSIE was launched with the main satellite HISAKI on September 14, 2013, and the first check-up data showed the successful charge-discharge cycling profiles of the LIC. However, the LIC has no longer been charged due to the malfunction of the PV panel since 1 month after the launch. Since then, as the best effort, the voltage trend of the LIC has been monitored to see whether long-term exposure to space environment has negative influence on the LIC. As of this writing, no malfunctioning trend has been observed, suggesting the LIC has still been performing well.

Author details

Masatoshi Uno[1*] and Akio Kukita[2]

*Address all correspondence to: masatoshi.uno.ee@vc.ibaraki.ac.jp

1 Ibaraki University, Hitachi, Japan

2 Japan Aerospace Exploration Agency, Sagamihara, Japan

References

[1] JCC. UPS-J EDLC Module [Internet]. Available from: http://www.jcc-foil.co.jp/cdg/products/index.html

[2] M. Uno and K. Tanaka. Accelerated charge-discharge cycling test and cycle life prediction model for supercapacitors in alternative battery applications. IEEE Transactions on Industry Electronics. 2012;59(12):4704–4712.

http://dx.doi.org/10.5772/64966

[3] M. Uno and A. Kukita. Cycle life evaluation based on accelerated aging testing for lithium-ion capacitors as alternative to rechargeable batteries. IEEE Transactions on Industry Electronics. 2016;63(3):1607–1617.

[4] M. Uno. Supercapacitor-Based Electrical Energy Storage System. InTech: Energy Storage in the Emerging Era of Smart Grids. 2011:21–44.

[5] X. Wang, Y. Sone, H. Naito, C. Yamada, G. Segami, and K. Kibe. Cycle-life testing of large-capacity lithium-ion cells in simulated satellite operation. Journal of Power Sources. 2006;161(1):594–600.

[6] M. Uno and K. Tanaka. Spacecraft electrical power system using lithium-ion capacitors. IEEE Transactions on Aerospace and Electronic Systems. 2013;49(1):175–188.

[7] M. Uno, K. Ogawa, Y. Takeda, Y. Sone, K. Tanaka, M. Mita, and H. Saito. Development and on-orbit operation of lithium-ion pouch battery for small scientific satellite "REIMEI". Journal of Power Sources. 2011;196(20):8755–8763.

Chapter 8

Interdigitated MEMS Supercapacitor for Powering Heart Pacemaker

Hafzaliza Erny Zainal Abidin, Azrul Azlan Hamzah, Jumril Yunas, Mohd Ambri Mohamed and Burhanuddin Yeop Majlis

Additional information is available at the end of the chapter

http://dx.doi.org/10.5772/65127

Abstract

Power MEMS can be defined as microelectromechanical systems for power generation and energy conversion. Energy harvesting has become an increasingly popular option for powering electronic devices as a long-lasting power source. Energy scavenging is defined as the process by which the energy is derived such as vibration, solar, wind, and thermal. Energy harvesting from the environment can prolong the life cycle and reduce the maintenance costs of electronic devices. Among the various sources of energy storage, Among the various of energy storage, supercapacitor has recently gained much interest in fields such as bioMEMS, biomedical implants and power electronic devices due to its advantages such as high power density, rapid charge and discharge and unlimited number of recharge cycles. In biomedical and bioMEMS systems, an energy storage device is needed to power other active biomedical devices within the system. For implantable devices such as a heart pacemaker, the power requirement is in the range of 30–100 μW. Microsupercapacitors play an important role in energy harvesting system, such as collecting energy from ambient energy sources. Human body is very resourceful in generating micropower in the form of heat dissipation, deformation of elastic tissue, and motion. Due to the advantages of MEMS energy harvesting system, the system can be use widely for biomedical implant devices, such as heart pacemakers and hearing aids, and can be used for a long time and without the need for battery replacement. In this work, planar and double-stacked interdigital electrode supercapacitor designs were modeled using Coventorware software. From simulation, it is observed that for planar structure, the specific capacitance is 0.22 mF/cm^{-2}, and for double-stacked structure specific capacitance can be increased to 0.48 mF/cm^{-2}. In terms of specific power, the planar structure produces 0.99 mW/cm^{-2}, and the double-stacked structure produces 2.18 mW/cm^{-2}. These results highlight the superiority of the double-stacked MEMS interdigital supercapacitor design compared with its planar counterpart in terms of charging capacity and electrical performance, thus making it favorable for powering heart pacemaker.

Keywords: energy scavenging, interdigital electrode microsupercapacitor, heart pacemaker, Coventorware

1. Introduction

Biomedical implantable devices usually required micro power sources with smaller dimensions and higher power density. The output power generated is in the range of nanowatt to microwatt. Power requirement for implantable devices for heart pacemaker is 30-100 μW [1, 2]. An energy generation system producing power within that range would qualify as a probable candidate to replace the conventional battery system. Typically, a Lithium-ion battery is widely used in biomedical implants such as pacemakers. However, batteries have many disadvantages such as short lifetime and contain a finite amount of depletable chemical energy [3]. Thus, patients using biomedical implants devices such as cardiac pacemakers have to replace the battery once every 5 years to 10 years. To avoid frequent battery replacement, a renewable energy generation and storage system could become a vital solution to infinitely empower biomedical devices, without the need for any powering unit replacement [4, 5].

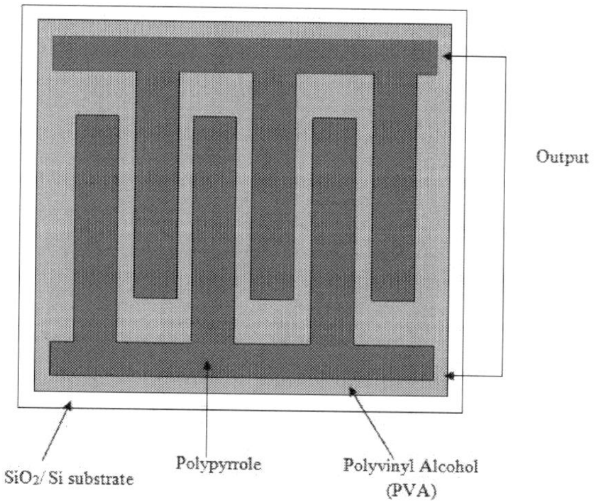

Figure 1. 2D view of planar interdigital electrode supercapacitor [23].

MEMS is one of the fundamental area that consists of mechanical, electrical, chemical, optical, and fluids engineering and integrated to the MEMS such as micropump,microvalve,microneedles and micronozzles [6–8]. Supercapacitor has high potential to replace conventional battery usage due to its high power density, rapid charge and discharge, long cycle life, high energy densities and unlimited number of recharge cycles. Miniaturized electrochemical capacitor or micro supercapacitor has promising capability to power small electronic devices. [9–11]. Supercapacitors have the capability to store up to 10,000 times more energy compared with its conventional counterparts [12, 13]. For electrode structure design of a supercapacitor, it can be classified into interdigital, sandwich, and roll structure [14]. Interdigital electrodes structure has their advantages such as low charging current, reduced solution resistance effects, and diffusion controllable current [15].

Conducting polymers have high capacitance and high conductivity with low cost compared to carbon as the electrode material for supercapacitor [16, 17]. Conducting polymers such as Polypyrrole (PPY), Polyaniline (PANI), and Polythiophene are mostly used as the electrode material in supercapacitors due to its advantages such as easy fabrication, low cost, high charge density, good conductivity, and more flexible [18–21]. In terms of electrolyte, Polyvinyl Alcohol (PVA) has been chosen due to its high charging density and low cost [22].

Planar interdigital electrode supercapacitor with polypyrrole (Ppy) – polyvinyl alcohol (PVA) as electrodes coating and electrolyte material respectively was first introduced by Wei Sun and Xuyuan Chen, having the advantage of high charging capacity due to its interdigital structure. In this structure, the supercapacitor consists of two Ppy coated nickel electrodes as current collectors and PVA as a solid-state electrolyte (**Figure 1**) [23].

In addition, carbon nano tube (CNT) can also be used as electrodes for the supercapacitor. In 2009, Jiang et al. [24] used carbon nano tube as supercapacitor electrodes and produced a specific capacitance of 428 $\mu F/cm^2$ and a power density of 0.28 mW/cm^2. In 2011, Shen et al. have used an activated carbon as part of a supercapacitor material to obtain a larger surface area. Beidaghi and Wang 2011 [25] also used carbon as supercapacitor electrodes to increase the surface area to be more effective but had to be improved in terms of ion diffusion through the thickness of the dielectric film for the electrodes.

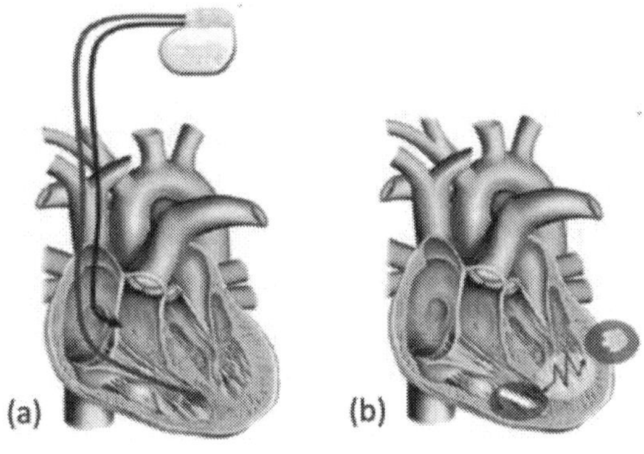

Figure 2. (a) Current pacemaker; (b) Future pacemaker using energy harvesting. Source: http://www.eetimes.com [26].

Microsupercapacitors play an important role in the energy harvesting system such ascollecting energy from ambient energy sources. In energy harvesting systems, it does not require any chemicals. In fact, the human body is very flexible in generating power from heat dissipation, deformation of elastic tissue, and the others. Due to its advantages as a energy harvesting system, it can be used widely for biomedical implant devices such as pacemakers and hearing aids and can be used for a long time and recharge without any replacement. **Figures 2** and **3** show examples of supercapacitor applications in MEMS devices. For a current pacemaker as depicted in **Figure 2**, the batteries are frequently changed and required a high cost for surgical operations compared with a pacemaker using energy harvesting that is more reliable and comfortable to the patients. For a normal pacemaker as shown in **Figure 2(a)**, it consists of two elements which are first, a pulse generator, placed under the skin in the chest, a battery, and the impulse control system, and second, a lead inserted directly in the heart through a vein, delivering the impulses [26]. **Figure 3** shows a small chip in the middle-ear cavity. The main components are the surgical implant, which is placed underneath the skin, and externally the audio processor that converts sound into electrical signals [26].

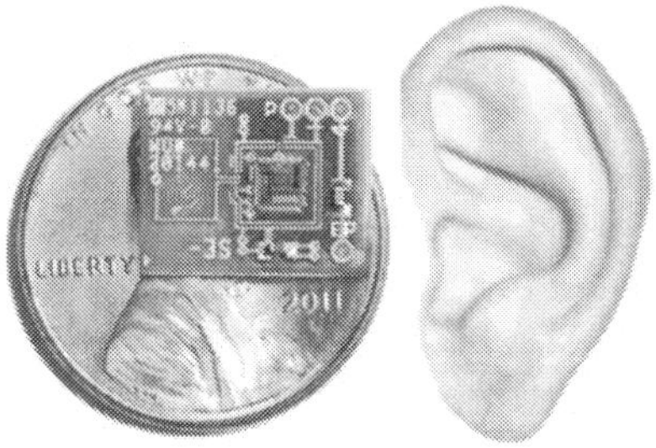

Figure 3. Small chip in the middle-ear cavity. Source: http://www.eetimes.com [26].

2. Design of planar interdigital electrode MEMS supercapacitor

In this research, planar interdigital electrode supercapacitor design was modeled using Coventorware ver.2008 via its process simulator tool. The software uses finite element analysis [27]. The planar supercapacitor structure consists of silicon (Si), silicon oxide (SiO_2), nickel (Ni), Polypyrrole (Ppy), and Polyvinyl Alcohol (PVA) layers. The structure was initiated by setting silicon as substrate 200 nm SiO_2 that was deposited on the silicon layer. Interdigital electrode structure was patterned using Ni on the SiO_2 layer. After that, the seed nickel layer was electroplated to construct high aspect ratio of nickel [28, 29]. The Ni electrodes were coated with Ppy to create the dielectric layer. PVA layer was fill deposited between the fingers to function as solid-state electrolyte. The process flow for modeling the structure of planar interdigital electrode supercapacitor is as shown in **Figure 4** and three-dimension structure of planar interdigital electrode supercapacitor is as shown in **Figure 5**.

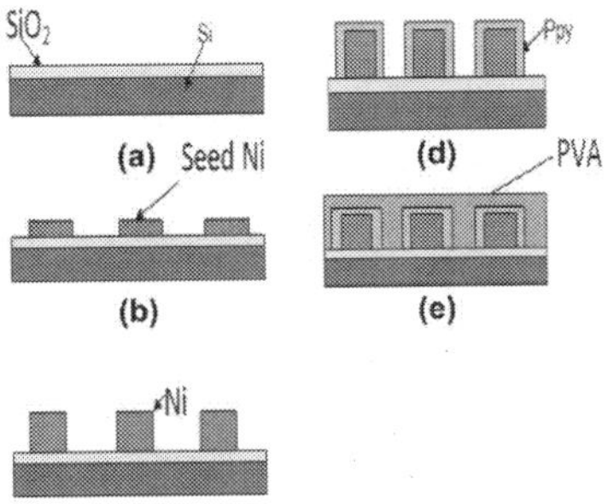

Figure 4. Process flow for modeling the double-stacked supercapacitor. (a) SiO_2 deposition, (b) Seed Ni deposition and patterning of 2D layout interdigital structure, (c) Ni growth of the interdigital structure, (d) Ppy coating on Ni, (e) PVA filling.

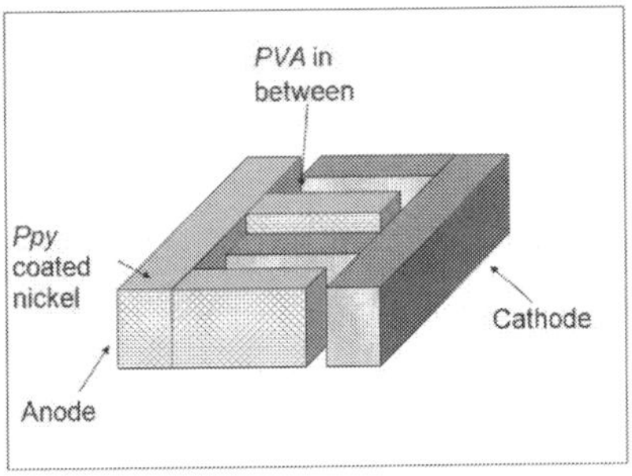

Figure 5. Structure of planar interdigital electrode supercapacitor.

3. Design of double-stacked interdigital electrode MEMS supercapacitor

Double-stacked supercapacitor design was modeled using Coventorware ver.2008 via its process simulator tool. The double-stacked supercapacitor structure consists of Si, SiO_2, Ni,

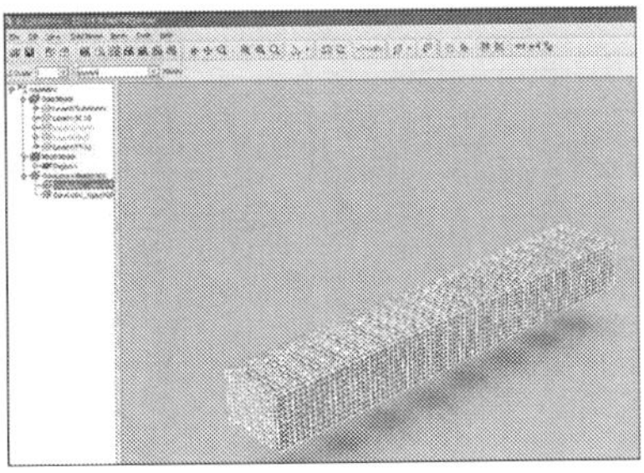

Figure 6. Meshing structure of planar interdigital electrode supercapacitor.

Ppy, and PVA layers. For a double-stacked structure, the mirrored cell from initial cell was developed by similar process steps. The electrodes would lie next to and atop of each other when the two cells are sandwiched together. The structure was initiated by setting silicon as substrate. 200 nm SiO2 was stack deposited on the silicon layer. Interdigital electrode structure was patterned using Ni on the SiO_2 layer. The Ni electrodes were coated with Ppy to create the dielectric layer. PVA layer was deposited between the fingers to function as a solid-state electrolyte (**Figures 6–9**).

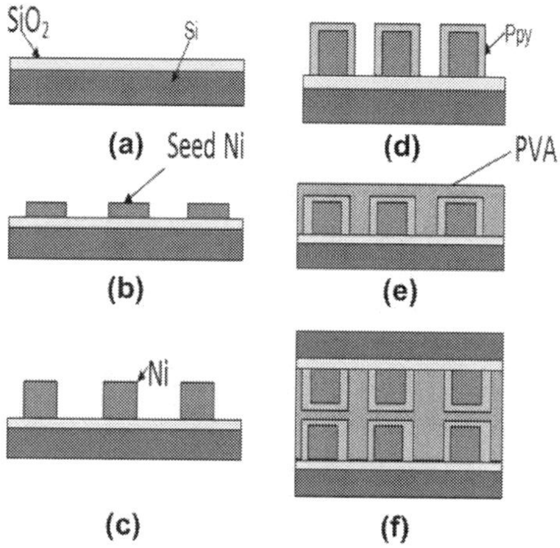

Figure 7. Process flow for modeling the double-stacked supercapacitor. (a) SiO2 deposition, (b) Seed Ni deposition and patterning of 2D layout interdigital structure, (c) Ni growth of the interdigital structure, (d) Ppy coating on Ni, (e) PVA filling and (f) Double-stack attachment.

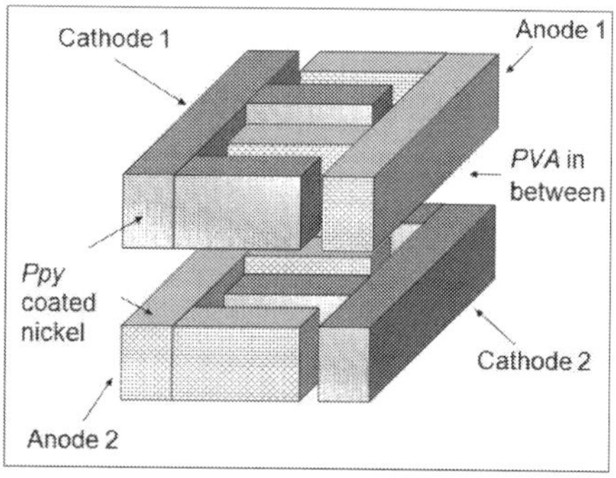

Figure 8. Structure of double-stacked interdigital supercapacitor with interwoven electrodes. Note that anode and cathode are always next to each other in bottom–bottom, top–top, and top–bottom electrode pairing configurations for the double-stacked design.

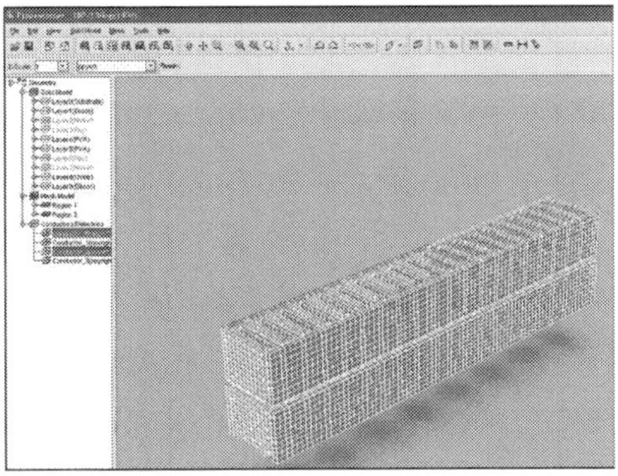

Figure 9. Meshing structure double-stacked interdigital electrode supercapacitor.

4. Electrical characteristics for planar interdigital electrode supercapacitor

Comsol Multiphysics ver.4.2 simulation tool was used to simulate the electrical characteristic of planar interdigital electrode design. Secondary current distribution and transport of diluted are chosen as the application modules [30]. Two dimensional time dependent cyclic voltammetry model was developed for the interdigital structure. Cyclic voltammetry measures current response when potential is applied to the working electrodes. For interdigital structure, the boundary conditions define the interface between the electrode and the electrolyte. The applied voltage range is set to −0.5 V to 0.5 V at the working electrode, and the current response at the working electrode can be measured. In the structure, Ppy-coated nickel is the current collector and PVA is the solid-state electrolyte. For the interdigital electrode, it consists of two electrode which is working electrode and counter electrode. The boundary conditions can be identify the interface between electrode and electrolyte. We set the boundaries at the counter electrode as ground and the electrode potential was applied at the working electrode boundaries based on 1[V]/Scan (t/1[s])-0.5[V] function [31, 32].

For cyclic voltammetry simulation, the bidirectional reaction can be formulated as in Ref. [3]:

$$A + e \overset{\rightarrow}{\underset{\leftarrow}{}} B \tag{1}$$

According to Eq. (2), for mass flow density of forward reaction N_f, it can be expressed as the rate constant k_f multiplied by concentration species during oxidation process C_O. For backward reaction, according to Eq. (3), for mass flow density of backward reaction N_b, it can be expressed as the rate constant k_b multiplied by concentration species during reduction process C_R.

$$N_f = k_f.Co \tag{2}$$

$$N_b = k_b . C_R \tag{3}$$

The rate constants of forward and backward reaction at the electrodes are described by the Butler-Volmer reaction kinetics

$$k_f = k_s e - \alpha \left(E - E_0\right) \frac{F}{RT} \tag{4}$$

$$k_b = k_s e(1 - \alpha)(E - E_0) \frac{F}{RT} \tag{5}$$

where k_s is the standard rate constant and α is the transfer coefficient. In symmetrical reactions, α has a value 0.5. E is the applied voltage on the electrode interface and E_0 has a value 0 V. For F, R, and T, it is a Faraday constant (96485 C/mol), molar gas constant (8.3144 J mol^{-1} K^{-1}), and temperature (298 K), respectively.

Additionally, the relationship for current, potential, and concentration can be summarized as follows:

$$\frac{i}{nFA} = k0\{c_O e[-\alpha\theta] - c_R e[(1-\alpha)\theta]\} \tag{6}$$

where $\theta = nF(E - E0)/RT$.

Initially, the Comsol simulation is simplified and performed on planar interdigital electrode structure as depicted in **Figure 8(a)**. It is observed that surface charge density is evenly distributed throughout the interdigital structures. As applied voltage between anode and cathode is slowly increased from 0 to 3 V, the charge density increases as more and more ions accumulate between the electrodes. A closer observation into the structure reveals that maximum charge density often occurs at the intersection of electrodes and its anchor, as depicted in **Figure 8(b)**. This is due to the proximity of the intersection point with the adjacent electrode, causing the charge to concentrate at the sharp edged points. The maximum charge density is as shown in **Figure 10(b)**.

Cyclic voltammetry portrays current response when potential is applied to the electrodes. In our case, the potential is applied at working electrode is between -0.5 V and 0.5 V, with the other electrode set as ground. The increasing voltage increases oxidation and reduction processes between the electrodes, driving current flow between working and counter electrodes. **Figure 11** shows that the oxidation and reductions occur within a voltage range of −0.5 and 0.5 V. For our design, ion flux reaches a peak within that region, resulting in maximum peak current (ip) of 6.0 A/m.

http://dx.doi.org/10.5772/65127

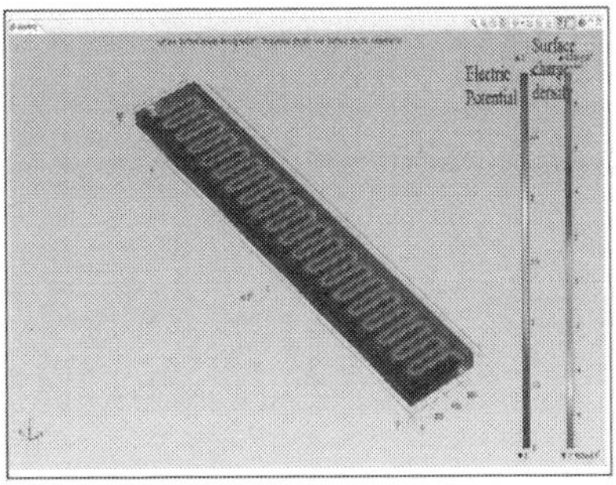

(a)

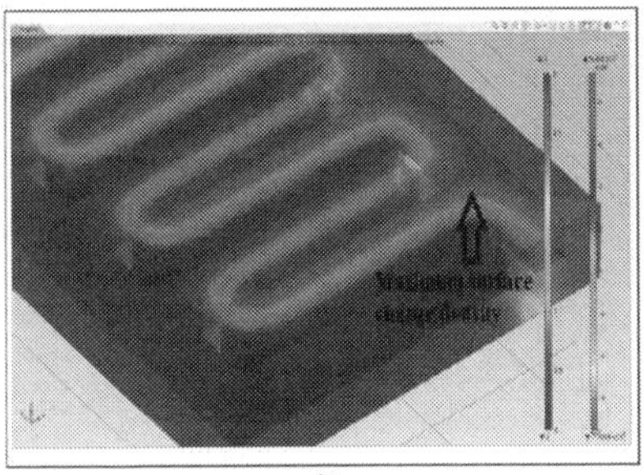

(b)

Figure 10. (a) Structure surface charge density; (b) Maximum surface charge density distributions of a planar supercapacitor with 20 finger pairs at an applied voltage of 3 V.

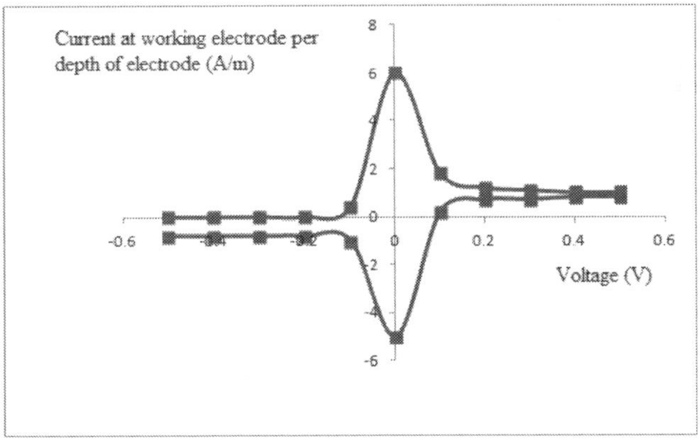

Figure 11. Cyclic voltammetry for planar interdigital electrode supercapacitor.

For a charge discharge of a microsupercapacitor, there are two techniques. One is the charging and discharging at a constant voltage to record current response with time, and the other is the charging discharging at a constant current to record the voltage response with time. In **Figure 12**, it is shown that when time is zero, the value of voltage is also zero. Discharging process for planar interdigital electrode occurred when the value of voltage decreased and was back to zero at 1 s.

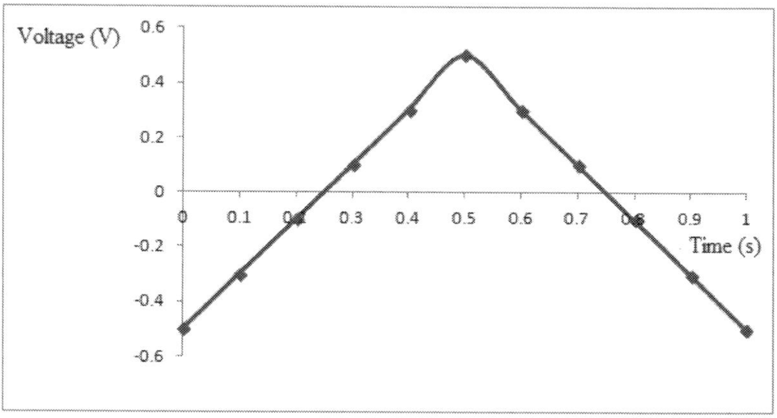

Figure 12. Charge discharge curve of planar electrode supercapacitor.

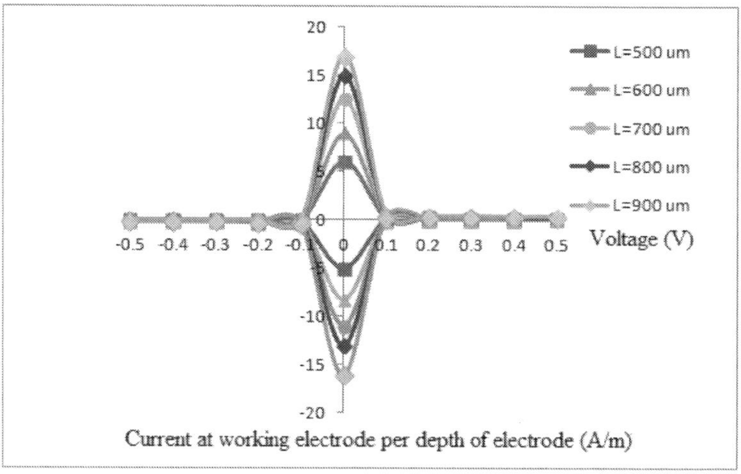

Figure 13. Cyclic voltammetry for various values of the length interdigital electrode.

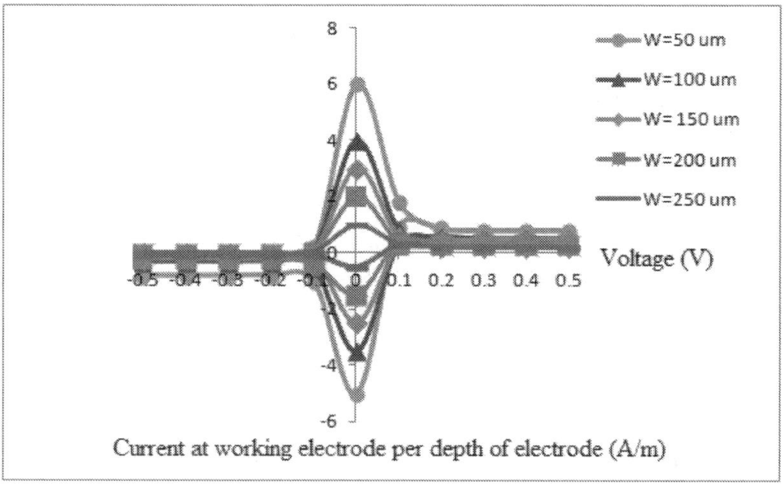

Figure 14. Cyclic voltammetry for various values of the width interdigital electrode.

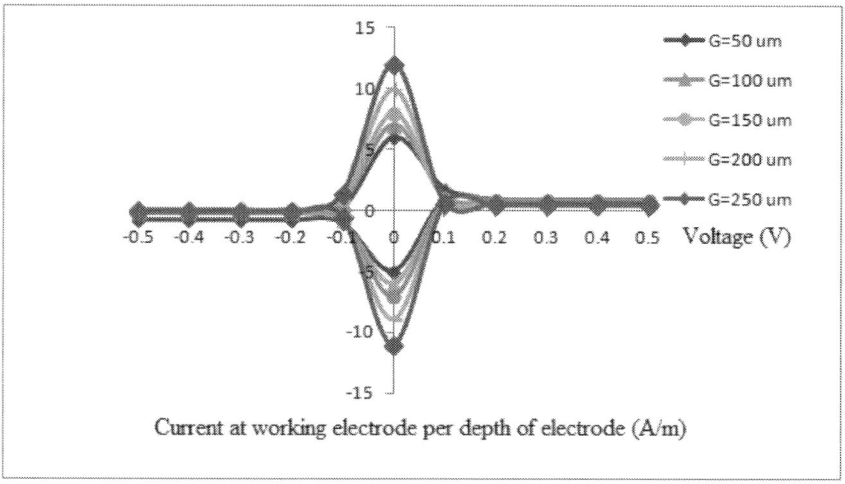

Figure 15. Cyclic voltammetry for various values of the gap interdigital electrode.

For the purposes of studying effects of design parameter changes, Comsol simulations were performed on the planar design with varying electrode length, width, and gap between the electrodes. Length of the interdigital electrodes are varied between 500 μm and 700 μm, with 700 μm being the maximum allowable length due to design limitation. From **Figure 13**, we can see the longer the electrode, the higher the current response. The maximum current response achieved is 1.6 A/m corresponding to electrode length of 700 μm. Increasing electrode length increases active surface area, which in turn enable a higher redox activity between the electrodes. Thus, increasing electrode length increases current generating mass transport activity on the cyclic voltammetry. Adversely, increasing electrode width from 50 μm to 250 μm produces the opposite effect as shown in **Figure 14**. From the figure, we can see that the smaller the width, the higher the current response. The maximum value of current response is 0.9 A/m, which occurs at electrode width of 50 μm. The reactions occurs at working electrode per depth of electrode data because the modelling is done in 2-D. The increasing value of gap increases space between the electrodes. This allows higher amount of active species to be present at the active interface. At the working electrode, the reduction and oxidation occurs more and more rapidly as more electrolyte are available with the increasing gap (**Figure 15**).

5. Electrical characteristics for double-stacked interdigital electrode supercapacitor

Cyclic voltammetry for the determination of a plane design double-digit supercapacitor between MEMS is the same as determining the design for cycle voltammetry for a coplanar. But there is a slight difference compared with the maximum current value for the design of the plane. **Figure 16** shows that the maximum flow for the cycle voltammetry multiple planar design is 13.2 A/m, which is more than twice the maximum current value relative to the plane designs. In the process of oxidation and reduction, equilibrium is reached and is limited by the voltage on the electrode surface. In the redox cycle, a decrease in current flow is from the effects of depletion. To complete the cycle, the voltage is not only directed to the front but also, on the other hand, within the range specified above. The curved shape also depends on the scan rate. Scan rate can be described as the speed at which the potential is varied. In addition,

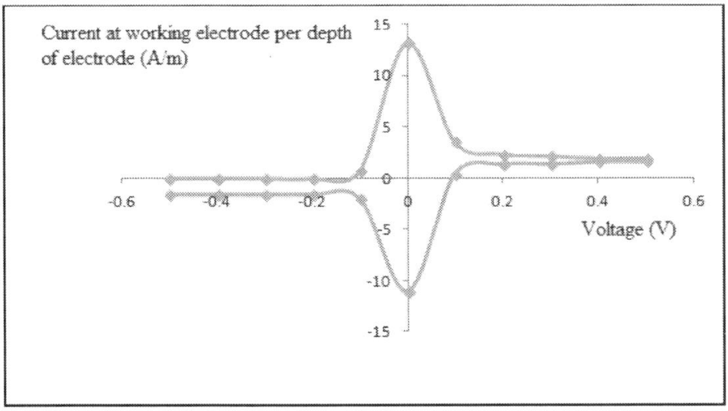

Figure 16. Cyclic voltammetry for double-stacked interdigital electrode supercapacitor.

Figure 17 shows the voltage response to the time indicated at 0.5 s, and the voltage reaches 1 V.

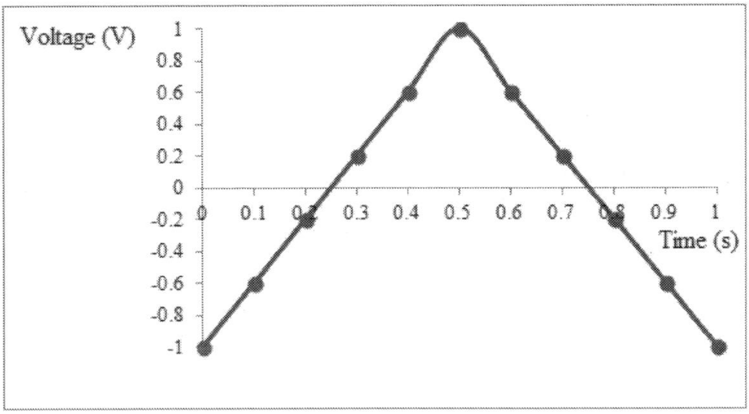

Figure 17. Charge discharge curve of double-stacked electrode supercapacitor.

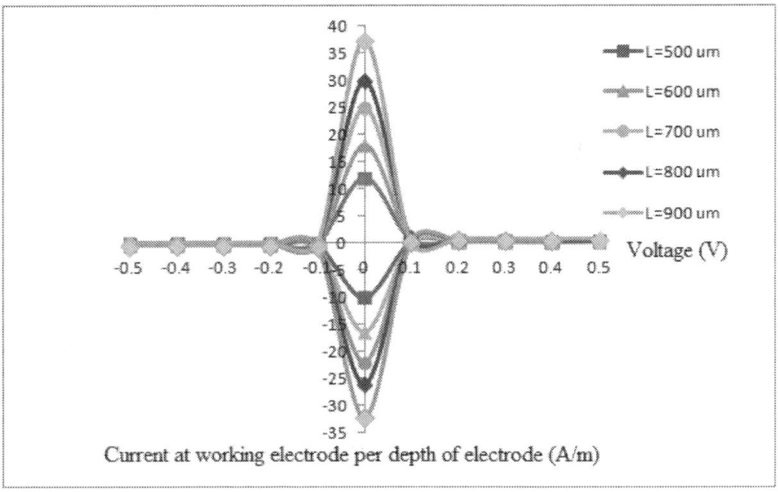

Figure 18. Cyclic voltammetry for various values of the length interdigital electrode.

Voltage is still in negative values until at t = 0.25 s. After that, the voltage increases until it reaches a maximum of 1.0 V at a value that is twice the value of the design of the plane at the time, t = 0.5 s, where the charging process occurs so as to achieve the maximum voltage. After reaching the maximum voltage, the discharge process occurs, whereby the voltage drop causes the charging rate to reduce back to a negative value of −1.0 V at a time $t = 1$ s.

For the purpose of studying the effects of changes in response to the current design of the multiple-plane and the plane, the value for the electrode length, width, and the gap between the electrodes was varied as the design of the plane. The length of the electrode between the digits are 500 and 900 m, where the length of the electrode between the digit at 900 m is the maximum length allowed for the design limits. According to **Figure 18**, it can be seen that the more the length of the electrode, the higher the reaction flow. The maximum current response was obtained at 37.4 A/m to 900 m-long electrode. Increasing the length of the electrode leads to an increase in active surface area, which in turn allows a higher redox activity to occur between the electrodes. Therefore, increasing the length of the electrodes improves the mass transport activity generated during the cycle voltammetry. High flow shows the life cycle of a higher supercapacitor.

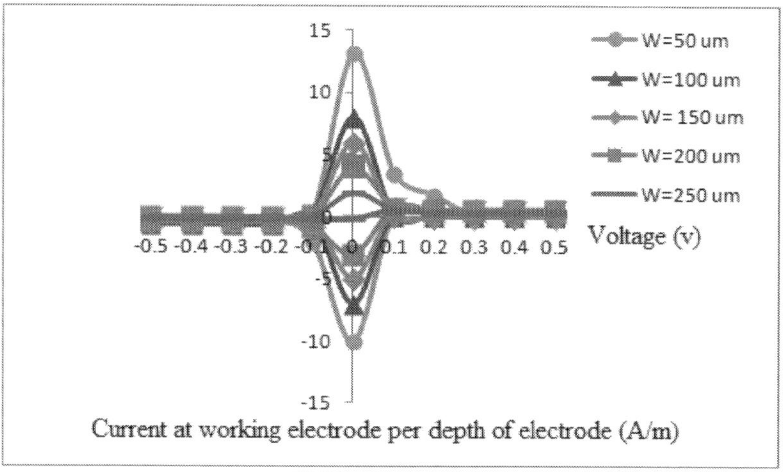

Figure 19. Cyclic voltammetry for various values of the width interdigital electrode.

According to **Figure 19**, we can see that the smaller the width of the electrode, the higher is the current response. The situation is similar to the design of the plane, and there is a difference in the maximum response current that is 13.2 A/m compared with 6.0 A/m for the design of the plane, which took place on the electrode width of 50 m. This phenomenon occurs because there is a balance between the surface area of the electrodes and the electrolyte. When the electrode width is increased, it will reduce the surface area of the electrolyte and also the amount of active species produced also decreasing.

For the gap between the electrodes to design multiple planes, the same value is used as in the design of the plane that is between 50 and 250 μm. In response to the voltage, resulting flows are plotted in **Figure 20**. Based on these figures, we can see that the greater the gap between the electrodes, the higher the current response. On the sidelines of the electrode 250 μm, the reaction flow reaches a maximum value of 26.4 A/m,which is doubled in value compared the design of the plane. This is because the number of cells for multiple planar design is twice that of the coplanar design that has 40 pairs of cells, so, with the increase in the number of cells, allowing for a number of active species of a higher existence on the active surface **Table 1** shows the comparison of parameters between planar and interdigital electrode supercapacitor design.

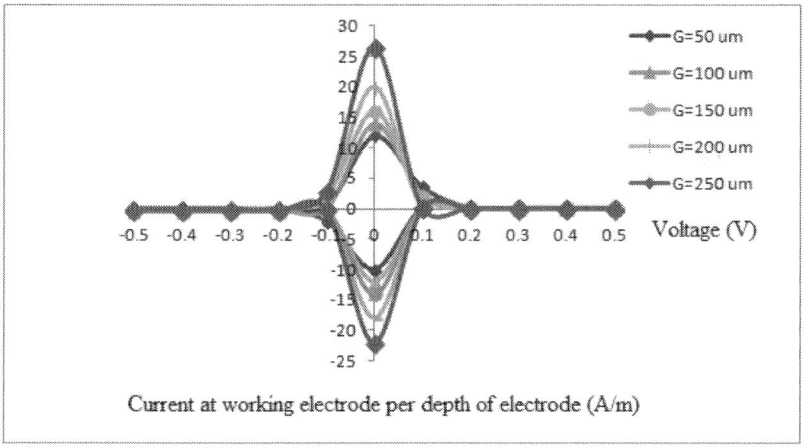

Figure 20. Cyclic voltammetry for various values of the width interdigital electrode.

Parameter	Planar	Double stacked
Number of fingers	20	Total 40 (20 top, 20 bottom)
Layout size (mm^2)	8.8	8.8
Thickness of cell (μm)	925	1950
Maximum charge (voltage input 3 V) (pC)	6.77	15.50
Specific capacitance (mF/cm^{-2})	0.22	0.48
Specific power (mW/cm^{-2})	0.99	2.18

Table 1. Comparison of parameters between planar and interdigital electrode supercapacitor design.

6. Conclusions

Energy harvesting system has a high potential as an alternative power especially in biomedical implant devices such as pacemakers. Due to the disadvantages such as short lifetime and containing a finite amount of depletable chemical energy of lithium-ion batteries, the patients using heart pacemaker have to replace the battery once every 5–10 years. To avoid any powering unit replacement, a renewable energy generation and storage system could become a vital solution. From this research, it can be concluded that double-stacked MEMS interdigital supercapacitor has same layout, only slightly thicker compared with the planar structure due to double stacking, but with much superior charging capacity. For both planar and double-stacked MEMS interdigital supercapacitor designs, the electrodes width, length, and gap between electrode fingers were fixed at 50, 500, and 50 μm, respectively. Furthermore, the planar and double-stacked MEMS interdigital supercapacitor designs were simulated using COMSOL ver.4.2a for electrical performance verification such as cyclic voltammetry and charge discharge performance. For cyclic voltammetry performance, applied voltage range is set to −0.5 to 0.5 V. For capacitance performance, it is observed that capacitance increases linearly with increasing number of cell, length of fingers, and width of fingers due to charge interactions among adjacent cells. The simulation results show that the planar structure has a charging capacity of 6.77 pC and the double-stacked structure has a charging capacity of 15.5 pC. Furthermore, for specific capacitance, it is observed that for planar structure is 0.22 mF/cm^{-2} and for double-stacked structure is 0.48 mF/cm^{-2}, while for specific power, the planar structure is 0.99 mW/cm^{-2} and for double-stacked structure is 2.18 mW/cm^{-2}. For charge discharge curve, it is observed that the curves are almost linear in the potential range. These results highlight the superiority of the double-stacked MEMS interdigital supercapacitor design compared with its planar counterpart in terms of charging capacity and electrical performance, thus making it favorable for powering heart pacemakers.

Author details

Hafzaliza Erny Zainal Abidin, Azrul Azlan Hamzah*, Jumril Yunas, Mohd Ambri Mohamed and Burhanuddin Yeop Majlis

*Address all correspondence to: azlanhamzah@ukm.edu.my

Institute of Microengineering and Nanoelectronics (IMEN), The National University of Malaysia, Bangi, Selangor, Malaysia

References

[1] Joo Hwan Sung, Se Joon Kim and Kun Hong. Fabrication of all solid state electrochemical microcapacitors. Journal of Power Sources. 2004;133(2):312–319.

[2] Venkateswara Sarma Mallela, V Ilankumaran and N Srinivasa Rao. Trends of cardiac pacemaker. Indian Pacing & Electrophysiology Journal. 4(4):201–212.

[3] Pei Hong Wang, Xu Hu Dai, Dong Ming Fang and Xiao Lin Zhao. Design, fabrication and performances of a new vibration based electromagnetic micro power generator. Microelectronics Journal. 2007;38(12):1175–1180.

[4] Craig L Schmidt and Paul M Skarstad. The future of lithium and lithium ion batteries in implantable medical devices. Journal of Power Sources. 2001;97–98:742–746.

[5] Hafzaliza Erny Zainal Abidin, Azrul Azlan Hamzah, Burhanuddin Yeop Majlis, Jumril Yunas, Norihan Abdul Hamid, Ummikalsom Abidin. Electrical characteristics of double stacked Ppy-PVA supercapacitor for powering biomedical MEMS devices. Microelectronic Engineering Dai. 2013;111:374–378.

[6] Noraini Marsi, Burhanuddin Yeop Majlis, Azrul Azlan Hamzah and Faisal Mohd Yasin Development of high temperature resistance of 500°C employing silicon carbide (3C-SiC) based MEMS pressure sensor. Microsystems Technologies. 2015;21:319–330.

[7] Jumril Yunas, Juliana Johari, Azrul Azlan Hamzah, Mimiwaty, Ille C Gebeshuber and Burhanuddin Yeop Majlis. Design and fabrication of MEMS micropumps using double sided etching. Journal of Microelectronics and Electronic Packaging. 2010;7:44–47.

[8] Azrul Azlan Hamzah, H E Zainal Abidin, B Yeop Majlis, M Mohd Nor, A Ismardi, G Sugandi, et al. Electrochemically deposited etched membranes with precisely sized micropores for biological fluids microfiltration. Journal Micromechanics and Microengineering. 2013;23:1–9.

[9] A Leela Mohana Reddy, F Estaline Amitha, Imran Jafri and S Ramaprabhu. Asymmetric flexible supercapacitor stack. Nanoscale Research Letters. 2008; 3(145):145–151.

[10] Hafzaliza Erny Zainal Abidin, Azrul Azlan Hamzah and Burhanuddin Yeop Majlis. Design of interdigital structured supercapacitor for powering biomedical devices. In: Regional Symposium on Micro and Nanoelectronics; Kota Kinabalu, Sabah, Malaysia. IEEE; 2011.

[11] Wei Sun, Ruilin Zheng and Xuyuan Chen. Symmetric redox supercapacitor based on micro fabrication with three dimensional polypyrrole electrodes. Journal of Power Sources. 2010;195(20):7120–7125.

[12] M Jayalakshmi and K Balasubramanian. Simple capacitors to supercapacitors – An overview. International Journal of Electrochemical Science. 2008;3:1196–1217.

[13] Maximilian Kaus, Julia Kowal and Dirk Uwe Sauer. Modelling the effects of charge redistribution during self discharge of supercapacitors. Electrochimica Acta. 2010;55(25):7516–7523.

[14] Caiwei Shen, Xiaohong Wang, Wenfeng Zhang and Feiyu Kang. A high performance three dimensional micro supercapacitor based on self supporting composite materials, Journal of Power Sources.2011:196(23):10465-10471.

[15] M Paeschke, U Wollenberger, C Kohler, T Lisec, U Schnakenberg and R Hintsche. Properties of interdigital electrode arrays with different geometries. Analytica Chimica Acta. 1995;2670(136):126–136.

[16] Meng Deng, Xi Yang, Musa Silke, Weiming Qiu, MingSheng Xu, Gustaaf Borghs et al. Electrochemical deposition of polypyrrole/graphene oxide composition on microelectrodes towards tuning the electrochemical properties of neural probes. Sensors and Actuators B. 2011;158(1):176–184.

[17] Marin S. Halper and James C Ellenbogen. Supercapacitor: A brief overviews. 2006:1–41.

[18] Hyuck Lee, Hyeongkeun Kim, Mi Suk Cho,Jaeboong Choi and Youngkwan Lee. Fabrication of polypyrrole (Ppy)/carbon nano tube (CNT) composite electrode on ceramic fabric for supercapacitor applications. Electrochimica Acta. 2011;56(22):7460–7466.

[19] R Ramya, R Sivasubramanian and M V Sangaranarayanan. Conducting polymers based electrochemical supercapacitor-progress and prospects. Electrochimica Acta. 2013;101:101–129.

[20] Farah Alvi, Manoj K Ram, Punya A Basnayaka, Elias Stefanakes, Yogi Goswami and Ashok Kumar. Graphene polyethylenedioxythiophene conducting polymer nanocomposite based supercapacitor. Electrochimica Acta. 2011;56(25):9406–9412.

[21] Mustafa Gullu and Deniz Yigit. A novel asymmetric pseudocapacitor based on poly (5,12-dihydrothieno[3,4:2,3] [1, 4] dioxocino[6,7-b] auinoxaline) coated graphite anode and poly(ethylenedioxythiophene)coated graphite cathodes. Electrochimica Acta. 2012;162(15–16):1434–1442.

[22] Graeme A Snook, Pon Kao and Adam S Best. Conducting polymer based supercapacitor devices and electrodes. Journal of Power Sources. 2011;196(1):1–12.

[23] Wei Sun and Xuyuan Chen. Fabrication and tests of a novel three dimensional micro-supercapacitor. Microelectronic Engineering. 2009;86(4–6):1307–1310.

[24] Y Q Jiang, Q Zhou and L Lin. Planar MEMS supercapacitor using CNT forests. In: 2009 IEEE 22nd International Conference on Micro Electromechanical Systems; IEEE; Sorrento, Italy. 2009. pp. 587–590.

[25] Majid Beidaghi and Chunlei Wang. Micro supercapacitor based on three dimensional interdigital polypyrrole/C-MEMS electrodes. Electrochimica Acta. 2011;56(25):9508–9514.

[26] Available from: http://www.embedded.com/print/4403423. 10 energy Harvesting solutions for 2012.

[27] Azrul Azlan Hamzah, Burhanuddin Yeop Majlis and Ibrahim Ahmad. Deflection analysis of epitaxially deposited polysilicon encapsulation for MEMS devices. In: IEEE,

http://dx.doi.org/10.5772/65127

editor. IEEE International Conference Semiconductor Electronics; Kuala Lumpur, Malaysia. 7–9 September 2004; 2004.

[28] Norazreen Abdul Aziz, B Bais, A A Hamzah and B Y Majlis. Characterization of HNA etchant for silicon microneedles array fabrication. In: IEEE 8th International Conference on Semiconductor Electronics; IEEE; Johor Bharu, Malaysia, 2008. pp. 203–206.

[29] N A Aziz, B Bais, A A Hamzah and B Y Majlis Optimization of HNA etching parameters to produce high aspect ratio solid silicon microneedles. Journal of Micromechanics and Microengineering. 2012;22(9)1-10.

[30] Hainan Wang and Laurent Pilon. Physical interpretation of cyclic voltammetry for measuring electric double layer capacitances. Electrochimica Acta. 2012; 64:130–139.

[31] Available from: https://www.comsol.com/community/exchange/203/. Robert Palumbo, Valparaiso University, Email: Robert.Palumbo@valpo.edu

[32] Alessandro Lavacchi, U Bardi, S Caporali, A Fossati and I Perissi. Cyclic voltammetry simulation at microelectrode arrays. In: Proceedings of Comsol Users Conference 2006; Bangalore, India. 2006.

Chapter 9

Power Management in Supercapacitor-Based Wireless Sensor Nodes

Hengzhao Yang and Ying Zhang

Additional information is available at the end of the chapter

http://dx.doi.org/10.5772/64987

Abstract

This chapter studies a power management problem for supercapacitor-based wireless sensor nodes with energy harvesting capabilities. A dependent task scheduling algorithm for nonpreemptable tasks with precedence constraints is developed. The modified first in first out (MFIFO) algorithm takes into account supercapacitor state and energy harvesting. Task precedence constraints are handled by defining a variable called task effective release time. Results show that the MFIFO algorithm improves the energy performance of the first in first out (FIFO) algorithm and maintains the timing performance at the same time.

Keywords: algorithm, energy harvesting, power management, supercapacitor charge redistribution, dependent task scheduling, wireless sensor network

1. Introduction

Wireless sensor networks have been developed for many applications. A wireless sensor network is composed of a large number of spatially distributed wireless sensor nodes. Wireless sensor nodes are usually powered by nonrechargeable batteries with limited capacity. Therefore, energy efficiency is a major concern. To maximize the network lifetime, various power management strategies have been proposed to minimize the energy consumption. In the meantime, numerous energy harvesting technologies have been developed to increase the energy income. Environmentally powered wireless sensor nodes usually need energy storage systems [1] to buffer the harvested energy. Typical energy storage systems include rechargeable batteries [2], supercapacitors [3–5], and hybrid systems [6, 7]. In general, rechargeable batteries have a larger capacity while supercapacitors have a much longer cycle life. The major drawback of supercapacitors is their high self-discharge rate. Supercapacitor characteristics must be taken into account to develop effective power management solutions. For instance, supercapacitor self-discharge is considered in Refs. [3, 6]. This is because the supercapacitor terminal voltage is a critical parameter in analyzing the power behavior of supercapacitor-based energy storage systems, and self-discharge results in voltage drop. Because of the significance of the voltage drop during self-discharge, this characteristic has been extensively examined [8–14].

While supercapacitor self-discharge leads to voltage drop, this characteristic cannot completely characterize the supercapacitor voltage behavior. In fact, the supercapacitor voltage may increase under the open circuit condition [15], which is due to the charge redistribution characteristic. A mechanism of the low ionic mobility in supercapacitor micropores is identified in Ref. [16]. The impact of charge redistribution on power management is qualitatively investigated in Ref. [17]. A detailed analysis of the voltage change during charge redistribution is performed in Ref. [18]. In Ref. [19], the modified earliest deadline first (MEDF) algorithm is developed for scheduling independent tasks.

This chapter extends the results in Refs. [17–19] and studies a new power management problem. Specifically, this chapter develops a modified first in first out (MFIFO) algorithm for scheduling tasks with precedence constraints in environmentally powered wireless sensor nodes that use supercapacitor-based energy storage systems. The MFIFO algorithm takes into account supercapacitor charge redistribution and energy harvesting. Task precedence constraints are handled by defining a variable called task effective release time. While the first in first out (FIFO) algorithm only considers the timing constraints of tasks, the MFIFO algorithm also considers the energy constraints.

The remainder of this chapter is organized as follows. Section 2 describes a system model for analyzing the power flow in wireless sensor nodes. Section 3 develops the MFIFO algorithm. Section 4 illustrates the implementation setup. A case study and extensive simulations are performed to evaluate the algorithm performance. These qualitative and quantitative results demonstrate that the MFIFO algorithm improves the energy performance of the FIFO algorithm while maintaining its timing performance. Section 5 concludes this chapter.

2. A power model for wireless sensor nodes

2.1. System model

This chapter adopts the wireless sensor node power model used in Ref. [19]. As shown in **Figure 1**, this model is composed of five modules: energy harvester, input power conditioning unit, energy buffer, output power conditioning unit, and energy user. Energy harvesters such as solar cells and piezoelectric films convert energy in other forms to electricity. Typically, an input power conditioning unit is needed to bridge the energy harvester and the energy buffer. For example, a solar-powered wireless sensor node usually includes a maximum power point tracker (MPPT). Energy buffers such as rechargeable batteries and supercapacitors are devices that store the harvested energy. An output power conditioning unit is often necessary to generate a suitable power supply for the energy user. DC-DC converters are commonly used modules to bridge the energy buffer and the energy user. Energy users are mainly RF transceivers, microcontrollers, and sensors.

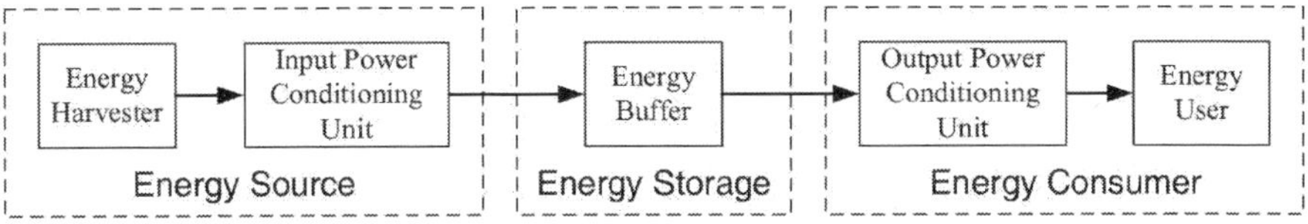

Figure 1. A power model for wireless sensor nodes.

The power model can be further abstracted to facilitate analyzing the power flow in wireless sensor nodes. As shown in **Figure 1**, the power model has three components: energy source, energy storage, and energy consumer. The energy source includes the energy harvester and the input power conditioning unit. For clarity, the energy buffer is referred to as the energy storage in this three-component model. The energy consumer combines the output power conditioning unit and the energy user. This abstracted model introduces two benefits. First, by separating energy buffers and power conditioning units, it is more convenient to study the impact of energy buffer characteristics on power management in wireless sensor nodes. Second, experiments with energy buffers can be readily designed and performed. The effects of input and output power conditioning units are taken into account in the process of designing the experiments.

2.2. Energy source model

The component models are shown in **Figure 2**. The energy source is modeled as a current pulse train. As shown in **Figure 2(a)**, each current pulse is characterized by three parameters: begin time B_S, duration D_S, and weight W_S, which is the current magnitude. It should be noted that the energy source pulse is the conditioned pulse that is actually injected into the energy storage system. For example, in a solar-powered sensor node, the current pulse conditioned by the MPPT and fed into the energy storage system is the current pulse described in this energy source model. By tuning these three parameters, energy source profiles with different characteristics such as time span and power level can be generated.

2.3. Energy storage model

The energy storage system is a single supercapacitor. **Figure 2(b)** shows the variable leakage resistance (VLR) model [10, 11, 17, 18] for supercapacitors, which is a simplified equivalent circuit model. In this model, the first branch has three components: resistor R_1, constant capacitor C_0, and voltage-dependent capacitor $K_V * V$. The total capacitance of the first branch is $C_1 = C_0 + K_V * V$. This branch models the voltage dependency of supercapacitor capacitance.

The second branch includes resistor R_2 and capacitor C_2. This branch models the charge redistribution behavior. The variable leakage resistor R_3 characterizes the time varying self-discharge.

In addition to the model parameters, the voltages across the capacitors in the VLR model are also critical to determine the supercapacitor state. The charge stored in a supercapacitor tends to redistribute among *RC* branches after a charging or discharging process because each branch

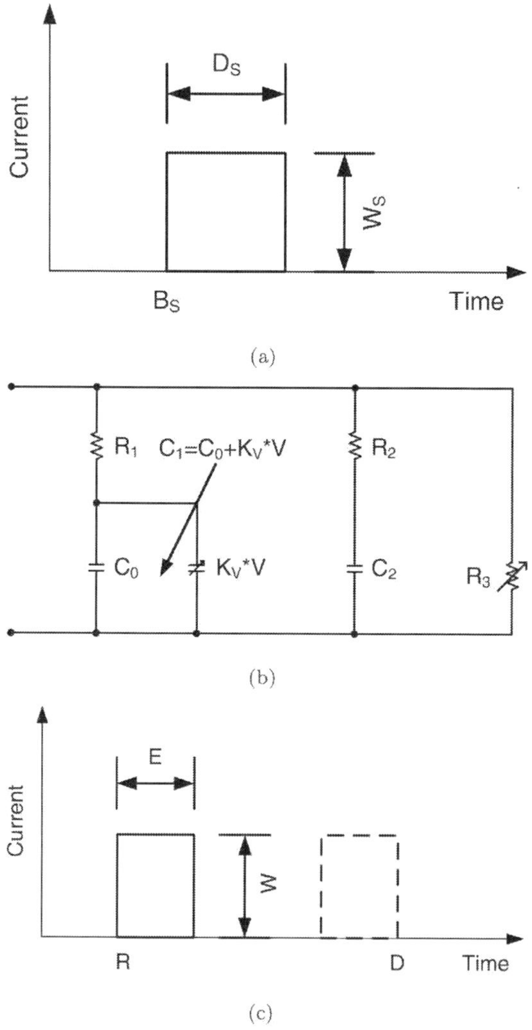

Figure 2. Wireless sensor node component models. (a) Energy source. (b) Energy storage. (c) Energy consumer.

http://dx.doi.org/10.5772/64987

has a different time constant. Charge redistribution is a transient response to the supercapacitor initial state, which is characterized by the initial voltages V_1 and V_2 across the capacitors C_1 and C_2. For example, if $V_1 > V_2$, the supercapacitor terminal voltage decreases with time because part of the charge stored in C_1 is transferred to C_2. Therefore, to characterize a supercapacitor, the initial voltages across the capacitors in the VLR model must be specified.

2.4. Energy consumer model

When the energy consumer initiates an event, the energy storage system is assigned a task. The energy consumer is therefore modeled as a current pulse train. As shown in **Figure 2(c)**, each task is defined by four parameters: release time *R*, execution time *E*, absolute deadline *D*, and weight *W*. The release time is the instant of time at which the task becomes available for execution. The execution time is the amount of time required to complete the execution. The absolute deadline is the instant of time by which the task execution is required to be completed. Release time, execution time, and deadline are temporal parameters defining the timing constraint of a task. The weight of a task is its current magnitude. This parameter defines the energy constraint of a task.

2.5. Task precedence constraint and effective release time

In this chapter, the tasks are assumed to be dependent and nonpreemptable. In addition to the four parameters (release time, execution time, deadline, and weight) used to characterize the task model, a task may also have precedence constraints. If tasks are constrained to execute in some order, they are said to have precedence constraints. The precedence constraints among tasks are specified using precedence relations [20]. A task T_p is a predecessor of another task T_q (and T_q a successor of T_p) if T_q cannot begin the execution until the execution of T_p completes. This fact is usually denoted by $T_p < T_q$. Two tasks are independent when neither $T_p < T_q$ nor $T_p > T_q$. A task with predecessors is ready for execution when the time is at or after its release time and executions of all the predecessors are completed. Without loss of generality, it is assumed that in this chapter a task may have no more than one predecessor or successor for simplicity.

The release times of tasks with precedence constraints are sometimes inconsistent with the precedence constraints, which means that the release time of a task may be later than that of its successor. **Figure 3** shows two tasks T_p and T_q using solid lines. If $T_p < T_q$, the release time of task T_q is earlier than the release time of T_p, which is not consistent with the precedence constraint. A parameter called the effective release time of a task is defined to deal with such inconsistency. The effective release time of a task without predecessor is equal to its release time. The effective release time of a task with predecessor is equal to the maximum value between its release time and the release time of its predecessor plus the execution time of its predecessor. For example, the effective release time ER_q of T_q is defined by Eq. 1 depending on whether there is a precedence constraint. As shown in **Figure 3**, the task T_q denoted by dashed lines shows its effective release time if $T_p < T_q$.

$$ER_q = \begin{cases} max(R_q, R_p + E_p), & T_p < T_q; \\ R_q, & otherwise. \end{cases} \quad (1)$$

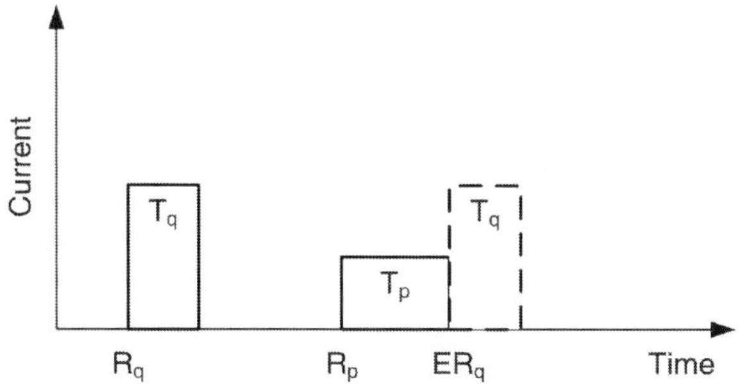

Figure 3. Definition of task effective release time.

3. MFIFO algorithm development

The MFIFO algorithm has three steps. First, create an initial schedule using the FIFO algorithm. This step takes care of the timing constraints and precedence constraints of tasks. Second, calculate the ready time adjustment margin based on the initial schedule. This margin determines how much delay is allowed if the ready time of the initial schedule is adjusted. Third, the ready time offset is determined based on supercapacitor state and energy harvesting. The start time of a task is the ready time plus the ready time offset.

3.1. Create an initial schedule using FIFO algorithm

The FIFO algorithm is used to create an initial schedule to ensure that the task precedence constraints are satisfied. The tasks are originally defined by the task set $T = T_i(R_i, E_i, D_i, W_i)$ and the precedence constraints $T_p < T_q$. The precedence constraints are transformed into timing constraints by defining the effective release times. A task is then characterized by four parameters: effective release time ER, execution time E, deadline D, and weight W. The task set is now $T_E = T_i(ER_i, E_i, D_i, W_i)$. The FIFO algorithm sorts the effective release times in the ascending order and determines the ready times of the tasks. The initial schedule is determined using Algorithm 1.

Algorithm 1: Create an initial schedule using FIFO algorithm.

Require: A set of N ready but not scheduled tasks: $T = T_i(R_i, E_i, D_i, W_i)$ and task precedence constraints: $T_p < T_q$.

for $i = 1 : N$ **do**
    if $T_p < T_q$ **then**
        $ER_q = max(R_q, R_p + E_p)$
    else
        $ER_q = R_q$
    end if
end for
Sort N tasks in the ascending order of their effective release times.
Current Time $T_C = 0$.
    for $i = 1 : N$ **do**
  Ready Time $A_i = max(T_C, ER_i)$
  Current Time $T_C = A_i + E_i$
  end for
  Algorithm output is initial schedule T_{FIFO} defined by task ready time A_i: $T_{FIFO} = T_i(A_i, E_i, D_i, W_i)$ and modified task set $T_E = T_i(ER_i, E_i, D_i, W_i)$.

3.2. MFIFO algorithm

Once the initial schedule is determined, the ready time adjustment margin and ready time offset are calculated using the algorithms for the second and third steps in the MEDF algorithm [19], respectively. In particular, the release times used in the MEDF algorithm should be replaced by the effective release times. The complete MFIFO algorithm is summarized in Algorithm 2. The inputs of this algorithm include a set of N ready but not scheduled tasks, $T = T_i(R_i, E_i, D_i, W_i)$; task precedence constraints, $T_p < T_q$; energy source model, $E_S(B_S, D_S, W_S)$; and supercapacitor initial state, $V_1(t = 0)$ and $V_2(t = 0)$. The MFIFO algorithm is a three-step process:

1. **Step 1:** Create an initial schedule using Algorithm 1. The input of this algorithm is the task set $T = T_i(R_i, E_i, D_i, W_i)$ and task precedence constraints $T_p < T_q$. The output is the initial schedule T_{FIFO} defined by task ready time A_i: $T_{FIFO} = T_i(A_i, E_i, D_i, W_i)$ and modified task set $T_E = T_i(ER_i, E_i, D_i, W_i)$.

2. **Step 2:** Calculate ready time adjustment margin of the initial schedule using the second algorithm in Ref. [19]. The inputs are the modified task set $T_E = T_i(ER_i, E_i, D_i, W_i)$ and the initial schedule $T_{FIFO} = T_i(A_i, E_i, D_i, W_i)$. The output is the task ready time adjustment margin M_i.

3. **Step 3:** Determine ready time offset of the initial schedule using the third algorithm in Ref. [19]. The inputs are the modified task set $T_E = T_i(ER_i, E_i, D_i, W_i)$, the initial schedule $T_{FIFO} = T_i(A_i, E_i, D_i, W_i)$, the ready time adjustment margin M_i, the energy source model

$E_S(B_S, D_S, W_S)$, and the supercapacitor initial state $V_1(t = 0)$ and $V_2(t = 0)$. The output is the modified schedule T_{MFIFO} defined by task start time S_i: $T_{MFIFO} = T_i(S_i, E_i, D_i, W_i)$.

Algorithm 2: MFIFO algorithm

Require: A set of N ready but not scheduled tasks: $T = T_i(R_i, E_i, D_i, W_i)$; task precedence constraints: $T_p < T_q$; energy source model: $E_S(B_S, D_S, W_S)$; and supercapacitor initial state: $V_1(t=0)$ and $V_2(t = 0)$.

1: Step 1: Create an initial schedule using Algorithm 1.

2: Input: task set T and task precedence constraints $T_p < T_q$.

3: Output: initial schedule T_{FIFO} and modified task set T_E.

4: Step 2: Calculate ready time adjustment margin of the initial schedule using the second algorithm in Ref. [19].

5: Input: modified task set T_E and initial schedule T_{FIFO}.

6: Output: task ready time adjustment margin M_i.

7: Step 3: Determine ready time offset of the initial schedule using the third algorithm in Ref. [19].

8: Input: modified task set T_E; initial schedule T_{FIFO}; ready time adjustment margin M_i; energy source model: $E_S(B_S, D_S, W_S)$; and supercapacitor initial state: $V_1(t = 0)$ and $V_2(t = 0)$.

9: Output: modified schedule T_{MFIFO}.

10: MFIFO Algorithm Output: modified schedule T_{MFIFO} defined by task start time S_i: $T_{MFIFO} = T_i(S_i, E_i, D_i, W_i)$.

4. MFIFO algorithm implementation and evaluation

4.1. Simulation setup

The MFIFO algorithm is implemented and evaluated using a simulation setup similar to the one used for the MEDF algorithm [19]. The energy source and energy storage models are exactly the same. The energy consumer model is modified. Each task set has six periodic tasks and each periodic task has five jobs. Therefore, each task set is composed of 30 tasks. The timing and energy parameters of a task are defined in the same way as the one used for the MEDF algorithm, too. The precedence constraints are assigned with controlled randomness. The six periodic tasks are partitioned into three groups. Each group consists of two periodic tasks. For convenience, the six periodic tasks are numbered as $\{P_1, P_2, \ldots, P_6\}$. The three groups are then $\{P_1, P_2\}$, $\{P_3, P_4\}$, and $\{P_5, P_6\}$. For each group, a job of the first periodic task is randomly selected as the predecessor of a job randomly selected from the second periodic task. Therefore, three pairs of precedence constraints are assigned for each task set. For example, **Figure 4** shows that in the first group $\{P_1, P_2\}$, the third job T_3 of the first periodic task P_1 is selected as the predecessor of the second job T_7 of the second periodic task P_2: $T_3 < T_7$. The MFIFO algorithm is evaluated in terms of the two metrics used for the MEDF algorithm: deadline miss rate and energy violation rate.

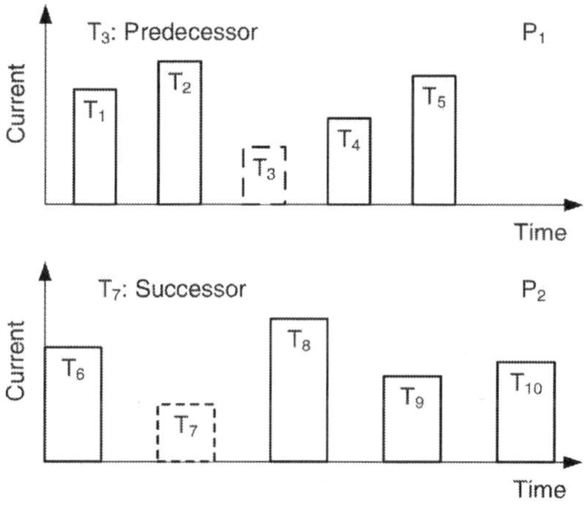

Figure 4. Assignment of precedence constraint.

4.2. An example

An example is used to illustrate the implementation and evaluation of the MFIFO algorithm. The simulation setup is adopted from Ref. [19], which is used to illustrate the MEDF algorithm implementation and evaluation. The supercapacitor initial state is $V_1(t=0) = V_2(t=0) = 1$ V. Two periodic tasks are used to define the task timing and energy constraints. The job T_2 from the first periodic task is selected as the predecessor of the job T_4 from the second periodic task. The precedence constraint is therefore $T_2 < T_4$. The effective release time of task T_4 is $ER_4 = max(R_4, R_2 + E_2) = 88$ s. The effective release times of the other five tasks are their release times. The task characteristics are listed in **Table 1** .

	T_1	T_2	T_3	T_4	T_5	T_6
Release time (R_i)	0	80	160	30	130	230
Effective release time (ER_i)	0	80	160	88	130	230
Execution time (E_i)	8	8	8	10	10	10
Deadline (D_i)	80	160	240	130	230	330
Weight (W_i)	35	30	40	42	37	33

Table 1. Tasks with precedence constraints to be scheduled.

The FIFO schedule determined using Algorithm 1 is shown in **Figure 5** . All the tasks are scheduled for execution at their effective release times. The task T_4 begins execution when its predecessor T_2 completes execution. The precedence constraint is satisfied. All the tasks respect their deadlines. The deadline miss rate is therefore α_{FIFO} = 0. The supercapacitor terminal voltage profile is shown in **Figure 6**. Two energy violations occur: T_1 and T_5. The minimum supercapacitor terminal voltages during the executions of the two tasks are 0.9670 and 0.9867 V, respectively. The energy violation rate is therefore β_{FIFO} = 2/6 = 0.333.

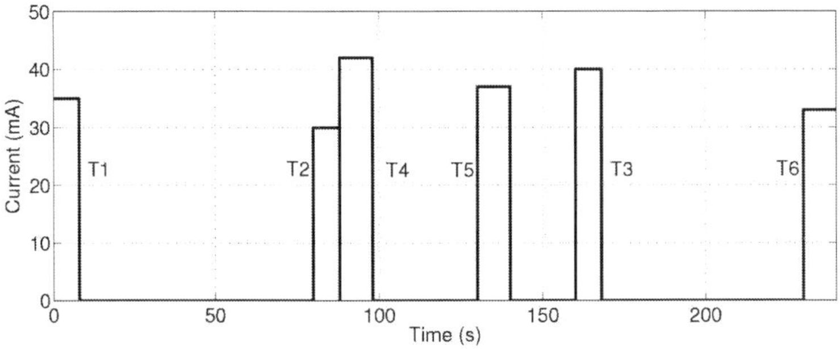

Figure 5. Task schedule determined using FIFO algorithm.

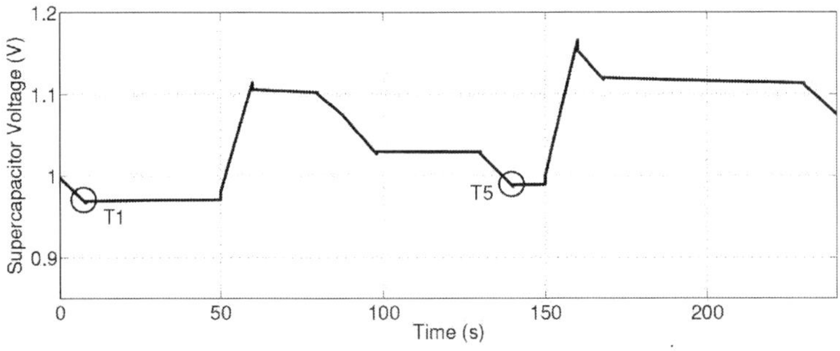

Figure 6. Supercapacitor terminal voltage profile of FIFO schedule.

The task ready time adjustment margin and task ready time offset results are listed in **Tables 2** and **3**, respectively. The task start time is then determined, and the MFIFO schedule is finalized. The MFIFO schedule is shown in **Figure 7**. Tasks T_1 and T_5 are postponed for execution. All the deadlines are respected. Therefore, the deadline miss rate is still α_{MFIFO} = 0. The MFIFO and FIFO algorithms have the same timing performance. The supercapacitor terminal voltage profile is shown in **Figure 8**. No energy violation occurs. The energy violation rate is β_{MFIFO} = 0. This example demonstrates that the MFIFO algorithm is better than the FIFO algorithm in terms of energy performance while maintaining the same timing performance.

http://dx.doi.org/10.5772/64987

FIFO schedule	T_1	T_2	T_4	T_5	T_3	T_6
FIFO schedule ready time (A_i)	0	80	88	130	160	230
Ready time flag ($F_i = A_i - ER_i$)	0	0	0	0	0	0
Maximum delay margin $MM_i = D_i - ER_i - E_i$	72	72	32	90	72	90
Available delay margin ($MA_i = MM_i - F_i$)	72	72	32	90	72	90
End time interval ($I_i = A_{i+1} - (A_i + E_i)$)	72	0	32	20	62	N/A
Ready time adjustment margin $M_i = min(MA_i, I_i)$	72	0	32	20	62	N/A

Table 2. FIFO schedule and ready time adjustment margin.

FIFO schedule	T_1	T_2	T_4	T_5	T_3	T_6
FIFO schedule ready time (A_i)	0	80	88	130	160	230
Ready time adjustment margin $M_i = min(MA_i, I_i)$	72	0	32	20	62	N/A
Execution time (E_i)	8	8	10	10	8	10
Latest end time $L_i = A_i + M_i + E_i$	80	88	130	160	230	N/A
$V_1(t = A_i)$	1	1.1005	1.0738	0.8825	1.1539	N/A
$V_2(t = A_i)$	1	1.0247	1.0287	1.0109	1.0352	N/A
If $V_1(t = A_i) > V_2(t = A_i)$	False	True	True	False	True	N/A
If $W_S(A_i < t < L_i) == 0$	False	True	True	False	True	N/A
Ready time offset (Δ_i)	72	0	0	20	0	N/A
MFIFO start time $S_i = A_i + \Delta_i$	72	80	88	150	160	230

Table 3. MFIFO schedule and ready time offset.

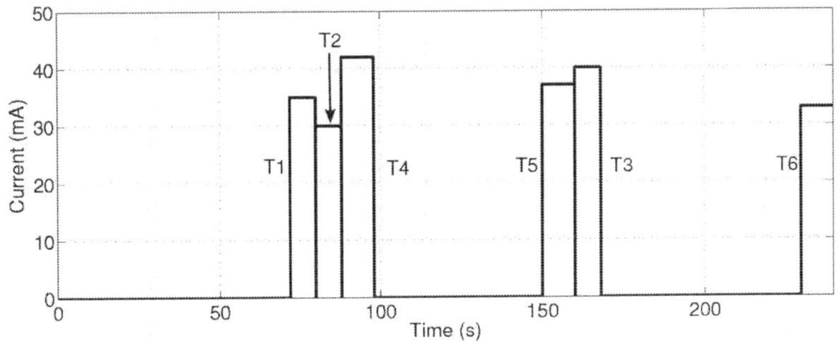

Figure 7. Task schedule determined using MFIFO algorithm.

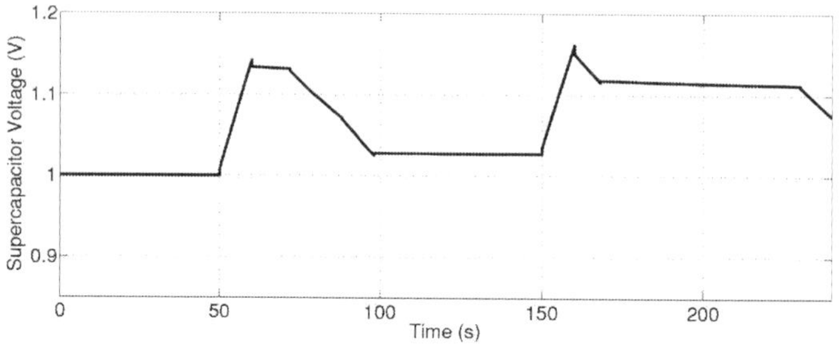

Figure 8. Supercapacitor terminal voltage profile of MFIFO schedule.

4.3. Evaluation results

The simulations are run for 200 times using the setup specified in Ref. [19]. The deadline miss rates and energy violation rates are recorded for the FIFO and MFIFO schedules. The obtained evaluation metrics are sorted in the ascending order and plotted. As shown in **Figure 9** , 35 out of the 200 simulation runs have zero deadline miss rates. The FIFO and MFIFO algorithms always have the same deadline miss rates. The timing and precedence constraints of the FIFO schedules are preserved in the MFIFO schedules. The energy violation rates are shown in **Figure 10** . For the FIFO algorithm, 96 out of the 200 simulation runs have an energy violation rate $\beta_{FIFO} = 1$. Among the 96 runs, the MFIFO schedules have an energy violation rate less than one for seven runs. Among the other 104 runs, the MFIFO algorithm results in an energy violation rate smaller than that of the FIFO algorithm for 81 runs. All together, the MFIFO schedules result in a smaller energy violation rate for 88 runs and a same energy violation rate for 112 runs. The simulation results verify that the MFIFO algorithm improves the energy performance of the FIFO algorithm and maintains the timing performance at the same time.

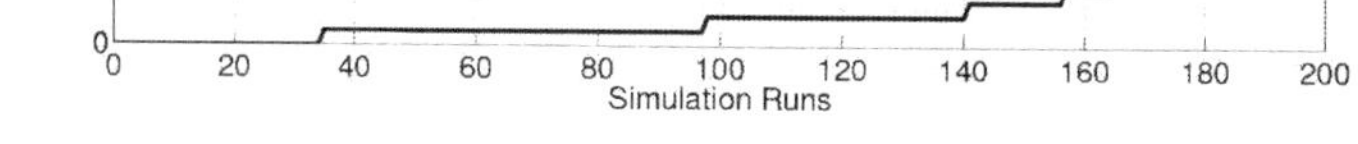

Figure 9. Deadline miss rates of FIFO and MFIFO algorithms.

The simulation setup is slightly modified to quantitatively compare the energy violation rates of the FIFO and MFIFO algorithms. The duty cycles of the six periodic tasks take the same value for each utilization. The utilization is $U = 6 * DC$, where DC is the duty cycle. The duty cycle increases from 0.02 to 0.1 with a step of 0.02. The utilization is swept from 0.12 to 0.6 with a step of 0.12. The other parameters of the tasks including periods, phases, weights, and precedence constraints are still defined using the setup specified in Ref. [19]. The simulations are run for 30 times for each utilization. **Figure 11** shows the calculated mean absolute percentage error (MAPE) values for the different utilizations. The MAPE decreases as the utilization increases. The average MAPE for the five utilizations is 12.1%. The MFIFO algorithm reduces the average energy violation rate of the FIFO algorithm by 12.1%. The MAPE is 25% for utilization $U = 0.12$. Like the MEDF algorithm, the MFIFO algorithm improves the energy violation rate more significantly if the sensor node operates with a relatively low duty cycle.

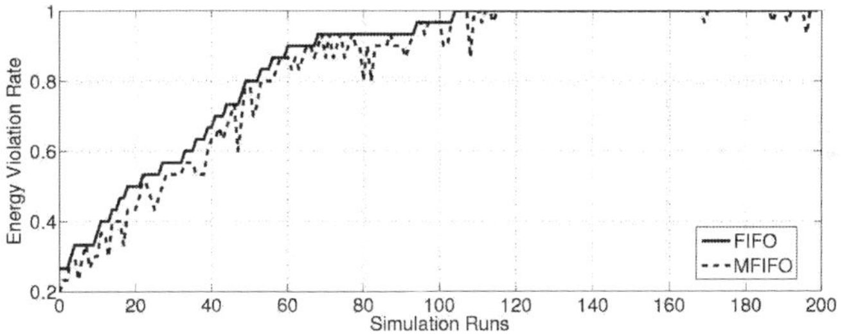

Figure 10. Energy violation rates of FIFO and MFIFO algorithms.

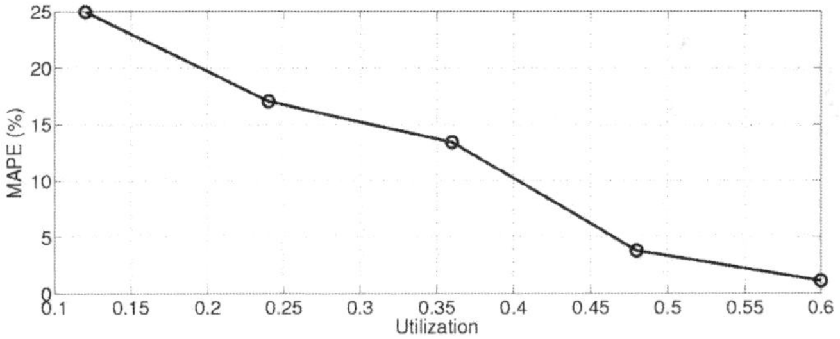

Figure 11. MAPE versus utilization for MFIFO algorithm.

5. Conclusion

This chapter proposes the MFIFO algorithm for nonpreemptable tasks with precedence constraints. The task precedence constraints are transformed into timing constraints by defining the effective release time of a task. The MFIFO algorithm takes into account the energy constraints of tasks in addition to the timing constraints. The MFIFO algorithm is implemented and evaluated. Simulation results show that the MFIFO algorithm improves the energy performance of the FIFO algorithm and maintains the timing performance at the same time.

Author details

Hengzhao Yang[1*] and Ying Zhang[2]

*Address all correspondence to: hengzhao.yang@csulb.edu

1 Department of Electrical Engineering, California State University Long Beach, Long Beach, CA, USA

2 School of Electrical and Computer Engineering, Georgia Institute of Technology, Atlanta, GA, USA

References

[1] S. Sudevalayam, P. Kulkarni, Energy harvesting sensor nodes: Survey and implications, IEEE Communications Surveys & Tutorials 13 (3) (2011) 443–461.

[2] C. Alippi, C. Galperti, An adaptive system for optimal solar energy harvesting in wireless sensor network nodes, IEEE Transactions on Circuits and Systems I, Regular Papers 55 (6) (2008) 1742–1750.

[3] T. Zhu, Z. Zhong, Y. Gu, T. He, Z.-L. Zhang, Leakage-aware energy synchronization for wireless sensor networks, in: MobiSys '09, 2009, pp. 319–332.

[4] D. Brunelli, C. Moser, L. Thiele, L. Benini, Design of a solar-harvesting circuit for batteryless embedded systems, IEEE Transactions on Circuits and Systems I: Regular Papers 56 (11) (2009) 2519–2528.

[5] F. Simjee, P. H. Chou, Everlast: Long-life, supercapacitor-operated wireless sensor node, in: ISLPED '06, 2006, pp. 197–202.

[6] X. Jiang, J. Polastre, D. Culler, Perpetual environmentally powered sensor networks, in: IPSN '05, 2005, pp. 463–468.

[7] H. Yang, Y. Zhang, Modeling and analysis of hybrid energy storage systems for wireless sensor networks, in: Proceedings of SPIE, Vol. 7647, 2010, pp. 76472U:1–76472U:10.

[8] B. E. Conway, W. Pell, T.-C. Liu, Diagnostic analyses for mechanisms of self-discharge of electrochemical capacitors and batteries, Journal of Power Sources 65 (1997) 53–59.

[9] Y. Diab, P. Venet, H. Gualous, G. Rojat, Self-discharge characterization and modeling of electrochemical capacitor used for power electronics applications, IEEE Transactions on Power Electronics 24 (2009) 510–517.

[10] Y. Zhang, H. Yang, Modeling and characterization of supercapacitors for wireless sensor network applications, Journal of Power Sources 196 (8) (2011) 4128–4135.

[11] H. Yang, Y. Zhang, Self-discharge analysis and characterization of supercapacitors for environmentally powered wireless sensor network applications, Journal of Power Sources 196 (20) (2011) 8866–8873.

[12] H. Yang, Y. Zhang, Evaluation of supercapacitor models for wireless sensor network applications, in: Proceedings of the 5th International Conference on Signal Processing and Communication Systems, 2011, pp. 1–6.

[13] H. Yang, Y. Zhang, Modeling and analysis of a solar powered wireless sensor node, in: Proceedings of the 2012 International Conference on Computing, Networking and Communications, 2012, pp. 970–974.

[14] H. Yang, Task scheduling in supercapacitor based environmentally powered wireless sensor nodes, Ph.D. thesis, Georgia Institute of Technology, 2013 Atlanta, GA, USA.

[15] M. Kaus, J. Kowal, D. U. Sauer, Modelling the effects of charge redistribution during self-discharge of supercapacitors, Electrochimica Acta 55 (25) (2010) 7516–7523.

[16] J. W. Graydon, M. Panjehshahi, D. W. Kirk, Charge redistribution and ionic mobility in the micropores of supercapacitors, Journal of Power Sources 245 (2014) 822–829.

[17] H. Yang, Y. Zhang, Analysis of supercapacitor energy loss for power management in environmentally powered wireless sensor nodes, IEEE Transactions on Power Electronics 28 (11) (2013) 5391–5403.

[18] H. Yang, Y. Zhang, A study of supercapacitor charge redistribution for applications in environmentally powered wireless sensor nodes, Journal of Power Sources 273 (2015) 223–236.

[19] H. Yang, Y. Zhang, A task scheduling algorithm based on supercapacitor charge redistribution and energy harvesting for wireless sensor nodes, Journal of Energy Storage 6 (2016) 186–194.

[20] J. W. S. Liu, Real-Time Systems, 1st Edition, Prentice Hall, 2000 Upper Saddle River, NJ, USA.